Ceramic Nanomaterials and Nanotechnology III

T0328409

Technical Resources

Journal of the American Ceramic Society

www.ceramicjournal.org

With the highest impact factor of any ceramics-specific journal, the *Journal of the American Ceramic Society* is the world's leading source of published research in ceramics and related materials sciences.

Contents include ceramic processing science; electric and dielectic properties; mechanical, thermal and chemical properties; microstructure and phase equilibria; and much more.

Journal of the American Ceramic Society is abstracted/indexed in Chemical Abstracts, Ceramic Abstracts, Cambridge Scientific, ISI's Web of Science, Science Citation Index, Chemistry Citation Index, Materials Science Citation Index, Reaction Citation Index, Current Contents/ Physical, Chemical and Earth Sciences, Current Contents/Engineering, Computing and Technology, plus more.

View abstracts of all content from 1997 through the current issue at no charge at www.ceramicjournal.org. Subscribers receive full-text access to online content.

Published monthly in print and online. Annual subscription runs from January through December. ISSN 0002-7820

International Journal of Applied Ceramic Technology

www.ceramics.org/act

Launched in January 2004, *International Journal of Applied Ceramic Technology* is a must read for engineers, scientists,and companies using or exploring the use of engineered ceramics in product and commercial applications.

Led by an editorial board of experts from industry, government and universities, *International Journal of Applied Ceramic Technology* is a peer-reviewed publication that provides the latest information on fuel cells, nanotechnology, ceramic armor, thermal and environmental barrier coatings, functional materials, ceramic matrix composites, biomaterials, and other cutting-edge topics.

Go to www.ceramics.org/act to see the current issue's table of contents listing state-of-the-art coverage of important topics by internationally recognized leaders.

Published quarterly. Annual subscription runs from January through December. ISSN 1546-542X

American Ceramic Society Bulletin

www.ceramicbulletin.org

The *American Ceramic Society Bulletin*, is a must-read publication devoted to current and emerging developments in materials, manufacturing processes, instrumentation, equipment, and systems impacting the global ceramics and glass industries.

The *Bulletin* is written primarily for key specifiers of products and services: researchers, engineers, other technical personnel and corporate managers involved in the research, development and manufacture of ceramic and glass products. Membership in The American Ceramic Society includes a subscription to the *Bulletin*, including online access.

Published monthly in print and online, the December issue includes the annual *ceramicSOURCE* company directory and buyer's guide. ISSN 0002-7812

Ceramic Engineering and Science Proceedings (CESP)

www.ceramics.org/cesp

Practical and effective solutions for manufacturing and processing issues are offered by industry experts. CESP includes five issues per year: Glass Problems, Whitewares & Materials, Advanced Ceramics and Composites, Porcelain Enamel. Annual subscription runs from January to December. ISSN 0196-6219

ACerS-NIST Phase Equilibria Diagrams CD-ROM Database Version 3.0

www.ceramics.org/phasecd

The ACerS-NIST Phase Equilibria Diagrams CD-ROM Database Version 3.0 contains more than 19,000 diagrams previously published in 20 phase volumes produced as part of the ACerS-NIST Phase Equilibria Diagrams Program: Volumes I through XIII; Annuals 91, 92 and 93; High Tc Superconductors I & II; Zirconium & Zirconia Systems; and Electronic Ceramics I. The CD-ROM includes full commentaries and interactive capabilities.

Ceramic Nanomaterials and Nanotechnology III

Ceramic Transactions Volume 159

*Proceedings of the 106th Annual Meeting
of The American Ceramic Society,
Indianapolis, Indiana, USA (2004)*

Editors
Song Wei Lu
Michael Z. Hu
Yury Gogotsi

Published by
The American Ceramic Society
PO Box 6136
Westerville, Ohio 43086-6136
www.ceramics.org

ISBN 1-57498-180-3

Contents

Synthesis and Functionalization of Nanoparticles

Nanostructured Membranes, Films, Coatings, and Self-Assembly

Processing and Characterization of Nanomaterials

Nanotubes and Nanorods

Environmental and Health Applications and the Future of Nanotechnology

Preface

This volume of Ceramic Transactions consists of papers presented at Symposium 6 on Nanostructured Materials and Nanotechnology, which was held during the 106th Annual Meeting of the American Ceramic Society (ACerS) in Indianapolis, Indiana, April 18-21, 2004. A total of twenty-nine papers are included in this volume plus a summary of a nanotechnology panel discussion. It divides into five chapters: Chapter I: Synthesis and Functionalization of Nanoparticles; Chapter II: Nanostructured Membranes, Films, Coatings, and Self-assembly; Chapter III: Processing and Characterization of Nanomaterials; Chapter IV: Nanotubes and Nanorods; and Chapter V: Environmental and Health Applications and the Future of Nanotechnology. This book is the third symposium proceedings titled "Ceramic Nanomaterials and Nanotechnology" from the unique nanotechnology symposium series during the annual meetings of the American Ceramic Society.

The symposium 6 followed three successful symposia on "Nanostructured Materials and Nanotechnology" during ACerS annual meetings in 2001, 2002, and 2003. The 2004 symposium consisted of eight sessions on ceramic nanomaterials, including Synthesis and Functionalization of Nanoparticles; Nanostructured Membranes, Films, Coatings, and Self-assembly; Dispersion and Sintering of Nanomaterials; Characterization Techniques for Nanomaterials; Nanotubes, Nanorods, Nanowires I: Carbon Nanotubes; Nanotubes, Nanorods, Nanowires II: Alternatives; Nanotechnology for the Future: Environment, Health, and National Security; and a panel discussion on nanotechnology: past, current, and future.

A total of 104 papers were presented in the symposium: 75 oral papers including 15 invited papers, and 28 poster papers, reflecting a significant increase of forty percentage from the same symposium a year ago. This well-organized and well-attended symposium was the largest symposium during the three-day conference. Authors came from seventeen countries and regions around the world, and from academia, national laboratories, industries, and government agencies, truly indicating the international impacts of this symposium and the broadening research activities of nanostructured materials and nanotechnology. Extensive collaborations between authors from different countries, different disciplines, and from academia, national laboratories, and industries have been strongly evidenced.

The symposium was sponsored by PPG Industries, Inc., and the Basic Science Division of the American Ceramic Society. The organizers sincerely thank these sponsors for their kind supports to ensure a very successful symposium. The editors would like to thank Greg Geiger at ACerS Headquarters for invaluable assistance in organizing the review process and coordinating the production of this volume of Ceramic Transactions. Finally, the editors thank all of the authors who contributed manuscripts to the proceedings or who assisted with reviewing manuscripts for this volume.

Editors

Song Wei Lu

Michael Hu

Yury Gogotsi

Synthesis and Functionalization of Nanoparticles

Liquid-Feed Flame Spray Pyrolysis of Single and Mixed Phase Mixed-Metal Oxide Nanopowders

R.M. Laine,* J. Marchal, S. Kim, J. Azurdia, and M. Kim
Contribution from the Department of Materials Science and Engineering,
University of Michigan, Ann Arbor, MI 48109-2136

ABSTRACT

Liquid-feed flame spray pyrolysis (LF-FSP) involves aerosolizing an alcohol solution containing alkoxide or carboxylate metalloorganic precursors with oxygen, igniting the aerosol to combust the alcohol fuel and precursor molecules, followed by rapid quenching at rates exceeding 500°C/sec. The resulting soot, or nanopowders typically consist of single crystal particles that are relatively unagglomerated with a composition identical to that in the original solution phase. The rapid cooling rates result in the production of nanopowders that are typically kinetic phase products. For example, LF-FSP provides access to nano δ^*-Al_2O_3 not α-Al_2O_3. Efforts to produce nanopowders along the Al_2O_3-TiO_2 tieline result in doping of Al^{3+} for Ti^{4+} ions inducing a phase transformation to the rutile crystal structure. LF-FSP of $Y_3Al_5O_{12}$ composition gives what appears to be a novel phase that is not YAG but does convert very easily to the YAG phase.

INTRODUCTION

Mixed metal-oxide (MMO) materials are used in areas ranging from photonics, to electronics, to catalysts, to health applications. Their widespread utility is based on their mechanical, electronic, photonic and catalytic properties. The potential of nanosized MMO powders ranges from the mundane (e.g. transparent sunblock cosmetics) to the highly sophisticated and novel (e.g. random, incoherent lasers). Unfortunately, there are currently no simple, general ways to make diverse MMO nanopowders with controlled chemical and phase composition. Thus, it is quite difficult to develop unified composition-structure-property relationships in such materials.

At the University of Michigan (UM), we have identified a new method of forming MMO nanopowders that combines simplicity with low-cost and exceptional generality. The process, called liquid-feed flame spray pyrolysis (LF-FSP), also represents a novel form of green manufacturing that could have considerable impact on traditional ceramics processing methods. The LF-FSP process appears to offer the means to make both single-phase MMOs and, in some instances, nanocomposite (mixed phase) materials in a single synthesis step.

In LF-FSP, alcohol solutions of low-cost chemical precursors are aerosolized with O_2, thereafter the aerosol is ignited and the resulting combustion process generates soot or oxide nanopowders with the exact composition of the precursor solution (to ppm), see Figure 1. The as-produced nanoparticles are predominantly high surface area, single crystal materials, often immediately ready for use. For example, LF-FSP has been used to produce nanocomposite powders of $CeO_{0.7}Zr_{0.3}O_2$ solid solution, phase segregated from and dispersed on α- or α-Al_2O_3. This MMO system, produced in one step, is identical to the system used commercially in three-way auto exhaust catalysts (TWCs). However, TWC washcoats are currently manufactured using 5-7 production steps. The catalytic activities of the LF-FSP materials are identical to those of commercial materials.

Figure 1. "Smaug" our LF-FSP reactor system, capable of producing up to 100 g/h of oxide nanopowders.

One of the primary difficulties facing those who wish to explore the potential of nanosized materials, e.g. metal oxides, is availability. Although there are numerous potential applications for nanopowders in areas ranging from luminescent materials,[1-9] to catalysts,[10] to nanocomposites,[12-14] the availability of large quantities of well-defined, high quality nanopowders is limited and costs can be high ($/g). Furthermore, there appear to be very few viable routes to large quantities of dispersible, chemically and phase pure mixed-metal oxide nanopowders. LF-FSP provides this opportunity as illustrated below where we discuss the synthesis of several examples of mixed-metal oxide nanopowders and some of their novel properties. First we briefly review methods of generating nanopowders.

<u>Typical methods of producing nanopowders</u>

Current methods of producing nanopowders can be broadly classified as liquid or gas phase processing. Liquid phase approaches include: sol-gel, precipitation, hydrothermal, sonochemical, shear-cavitation,[9] or electrochemical processing.[15-19] Gas phase approaches include:[11] spray pyrolysis, evaporative metal, flame spray pyrolysis, laser pyrolysis, or CVD methods.[20-29]

Of these methods, spray pyrolysis,[11,19] metal evaporation,[12,26] and flame spray pyrolysis[10,27] are the primary commercial methods of making nanopowders with the latter two offering the best control of particle size, composition and morphology. Historically, FSP is the primary commercial method of making large quantities of ultrafine powders (100-250 nm) and nanopowders (1-100 nm). Ultrafine and nanotitania, and fumed (nano) silica are produced in thousand ton/yr quantities by combusting volatile $TiCl_4$ or $SiCl_4$, in a H_2/O_2 flow. Combustion (> 1000°C) generates nanopowders as a ceramic soot, and chlorine and HCl gases as byproducts.[10,27] Although these byproduct gases are easily removed; they are toxic, corrosive pollutants as are the starting metal chlorides. The resulting nanopowders (e.g. titania) are not particularly different from those produced by metal evaporation or LF-FSP of metal alkoxides.[30,31]

Our interest in exploring LF-FSP derives from the discovery of methods of making very inexpensive single and mixed-metal metalloorganic precursors as illustrated in reactions (1)-(4):

$$SiO_2 + N(CH_2CH_2OH)_3 \xrightarrow[\text{x's } HOCH_2CH_2OH]{200°C/-H_2O} \text{silatrane}$$

(1)

$$Al(OH)_3 + N(CH_2CH_2OH)_3 \xrightarrow[]{200°C/EG/-H_2O} \text{alumatrane}$$

(2)

$$Mg(OH)_2 + 2Al(OH)_3 + 3\ TEAH_3 \xrightarrow[\text{x's EG}]{200°C/-H_2O}$$

(3)

$$SrO + 2SiO_2 + 2Al(OH)_3 + 6TEAH_3 \xrightarrow[\text{x's EG}]{200°C/-H_2O}$$

(4)

The precursors shown in reactions (1)-(4) are molecular species; but some precursors are simple homogeneous mixtures. The mullite precursor is a homogeneous 6:2 alumatrane: silatrane EtOH solution.

The Figure 1 FSP apparatus "Smaug" was constructed in our laboratories and is a tenth generation device. It combusts EtOH aerosols containing 1-20 wt. % ceramic (as precursor) with the optimum (for viscosity reasons) being 5-15 wt % depending on the precursor. CeO_2/ZrO_2 precursor solutions at 20 wt.% oxide can be "shot" quite easily (>100 g/h). Flame temperatures are ≤ 2000 °C based on down-stream measurements and calculations. The temperature drops rapidly (≈ 1400°C) to 400-300°C about 1 m from the flame, just beyond the main chamber (20 cm dia.).

Here we briefly describe selected materials produced by LF-FSP beginning with Table 2.

FSP of single metal n-oxides.

TiO_2. In-depth studies of nano-TiO_2 made by LF-FSP compare favorably with Degussa P-25 nano-TiO_2 and metal evaporation derived nanoTiO_2 (Nanophase). These nanoTiO_2s have similar specific surface areas (SSAs), particle sizes and are all approximately 90% anatase, 10 % rutile.

Al_2O_3. LF-FSP of various precursors including alumatrane, reaction (2), give the products such as shown in Figure 2. Note that the $AlCl_3$ and $Al(NO_3)_3$ precursors provide both nanopowders and mixtures of micron sized particles with some nanosized powders.

Table 1. Typical powders produced by LF-FSP at the University of Michigan.

Powder	Density(g/cc)	F(nm)[*,**,†]	Surface Area(m²/g)[*]	Remarks
TiO_2	--	60-20	30-80	(90-10 anatase-rutile)[†]
CeO_2	7.047-7.104	30-90	10-15	single & twinned crystals
Al_2O_3	..	40-20	50-80	δ-Al_2O_3
CaO	..	40-20	50-80	some $CaCO_3$ and $Ca(OH)_2$
MgO	..	40-20	50-80	trace $MgCO_3$ and $Mg(OH)_2$
CeO_2/ZrO_2		20-40	30-50	10/90 to 90/10 cubic
MgO Al_2O_3	--	50-30	40-60	spinel, faceted single xtals
$3Al_2O_3\ 2SiO_2$	2.825	70-80	40-50	mullite @ 1000°C
β"-alumina	$Na_{1.67}Al_{10.67}Li_{0.33}O_{16.33}$		30-50	partially crystalline
$SrO\ Al_2O_3\ 2SiO_2$		50-30	40-60	SAS, amorphous as shot

[*]BET, [**]Estimated from surface area/mass, [***,†]XRD Debye-Scherrer calculations. and/ TEM.

Figure 2. SEMs of $AlCl_3$ (a) and (b) $Al(NO_3)_3$ derived δ*-alumina powders

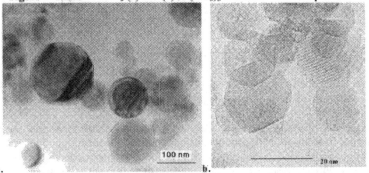

a.   b.

Figure 3. TEMs of alumatrane derived δ*-alumina (a) 5 wt. % solution, (b) 1 wt. % solution.

Although LF-FSP flame temperatures reach nearly 2000°C, the major phase observed for all conditions whether the particles are small or large or even hollow (Figure **2b**), is the d-phase. We have proposed that this is a consequence of the high water contents in the flame; however, even the Nanophase alumina powders are also d phase.[32] An alternative explanation is that the surface curvature for nanoalumina powders imparts such a high surface energy that the phase transformation from a density of 3.5 g/cc for d to 4.0 g/cc for a is not favorable.

These alumina precursors are the basis for a number of other materials. If we "shoot" aluma-trane with the titanium analog of silatrane (titanatrane) both of which are mutually soluble in EtOH, then we can explore the $(TiO_2)_x(Al_2O_3)_{1-x}$ phase space, which we have recently done with the idea of determining what could be used to provide optimal photocatalytic activity with good-to-excellent wetting.[33] As seen in Figure 4, we can make many different compositions very easily.

Perhaps the most important observation here is that the addition of just 15 mol % alumina forces a transformation from primarily anatase phase to rutile. This transformation appears to be driven by replacement of Ti^{4+} ions with Al^{3+} ions and the creation of oxygen vacancies making this material equivalent to a highly defective titania phase. The important implication is that the band gap of this material is not the same as that of rutile and may offer different photocata-lytic properties.

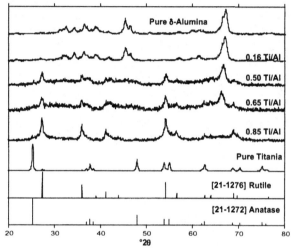

Figure 4. LF-FSP produced nanopowders in the $(TiO_2)_x(Al_2O_3)_{1-x}$ phase space. Average particle sizes are 2-40 nm with surface areas of 45-60 m^2/g.

$Y_3Al_5O_{12}$ Efforts to develop yttrium aluminum garnet nanopowders led to materials that have the correct composition but are not YAG.[34] Thus, the nano $Y_3Al_5O_{12}$ powders shown in Figure **5** were produced using an yttrium proprionate/aluminum acetylacetonate precursor. The powders are uniform on a micron length scale showing no large particles. At the nanometer length scale, they are also quite uniform with average particle sizes of 20 nm as shown in Figure **6**.

XRD powder pattens of samples of these powders reveal an apparently new phase with the $Y_3Al_5O_{12}$ composition that is an hexagonal perovskite. Sets of powders with the $Y_3Al_5O_{12}$ com-position from different precrusors were heated at 10°C/min and the transformation activation en-ergy was determined as shown in Table 2. The activation energy for this process is very low compared with the crystallization of YAG from glass and reflects the fact that the new phase is 30% denser than YAG which drives the transformation process.

Table 2. YAG starting formation temperature and E_a.*

Precursor	1	2	3	4	5	6	7	8	9
YAG formation (°C)	1280	1270	1060	1050	960	1100	1110	1100	1075
E_a (kJ/mol)	210	208	210	280	86	106	166	100	96

*The formation starting temperature was obtained at a heating rate of 10°C/min.

Figure 5. SEM micrograph of powders made from $Y(O_2CCH_2CH_3)_3/Al(O_2C_5H_8)_3$

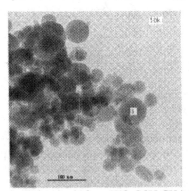

Figure 6. TEM micrograph of powders made from $Y(O_2CCH_2CH_3)_3/Al(O_2C_5H_8)_3$

As we will report at a later date, these nanopowders sinter much better than micron size YAG powders and we can achieve complete densification at 1400 °C in 6-8 h with final grain sizes of <200 nm with minimal processing. Thus, these materials offer potential for both structural applications and for transparent polycrystalline lasers.

In summary, LF-FSP provides access to a wide variety of well known materials as well as materials that are kinetic phases rather than thermodynamic phases. Consequently, LF-FSP also offers the opportunity to develop materials with very novel properties for many different applications.

ACKNOWLEDGEMENTS.

We would like to thank AFOSR (F49620-03-1-0389) and the DSO National Laboratories of Singapore, and Hyundai-Kia Motors Co. for support of this work.

REFERENCES
1. H.W. Leverence. An Introduction to Luminescence of Solids, Dover Publications, N.Y. 1968
2. 1968 Cornell Symp./APS. Preparation and Characteristics of Solid Luminescent Materials, J. Wiley & Sons, N.Y. 1968
3. D. Curie, Luminescence in Crystals, New York- J. Wiley & Sons, London- Methuen, 1963
4. a.F. Auzel, Proc. IEEE **61**, 758-86 (1973). b. F. Ostermayer, Met. Trans. **2**, 747-55 (1971).
5. G.E. Jabbour, J.E. Shaheen, M. Morrell, B. Kippelen, N. Armstrong, N.P Peyghambarian, Opt. and Photonics News, Apr. 1999, p. 25-27.
6. See for example the Special Issue of Synthetic Metals, **91**, (1997) on OLEDs.
7. L.L. Beecroft, C.K.Ober, Chem. Mater. **9**, 1302-17 (1997).
8. B.M. Tissue, Chem. Mater. **10**, 2037-41 (1998).
9. a. Advanced Catalysts and Nanostructured Materials: Modern Synthetic Methods, W.R. Moser, ed. Academic Press, San Diego, 1996. b. Ultra-fine Particles: Exploratory Science and Technology, C. Hayashi, R. Uyeda, A. Tasaki, eds. Noyes Publication, Westwood, NJ. 1997. c. The Journal Nanostructured Materials, Pergammon Press.
10. K.A. Klusters, S.E. Pratsinis, Powder Tech. **82**, 79-91 (1995).
11. A. Gurav, T. Kodas, T. Pluym, Y. Xiong, Aerosol Sci. Tech. **19**, 411-52 (1993).
12. R.W. Siegel, S. Ramasamy, H. Hahn, L. Zongquan, L. Ting, R. Gronsky, J. Mater. Res. (1988) **3**, 1367-72.
13. M. Gallas, B. Hockey, A. Pechenik, G. Piermarini, J. Am. Ceram. Soc. **77**, 2107-12 (1994).
14. W.H. Rhodes, in Phase Diagrams in Advanced Ceramics, A.M. Alper, ed., Academic Press, N.Y., N.Y. 1995 pp 1-39.
15. B. Alken, W.P. Hsu, E. Matijevic, J. Am. Chem. Soc. **71**, 845-53.
16. D. Sordelet, M. Akinc, J. Coll. and Interface Sci. **122**, 47-59 (1988).
17. M.E. Labib, J.H. Thomas, J. Electrochem. Solid-State Sci. and Tech. **134**, 3182-6 (1987) and references therein.
18. T. Kasuga, M. Hiramatsu, M. Hirano, A. Hoson, J. Mater. Res. **12**, 607-9 (1997).
19. A. Lopez, F.J. Lazaro, J.L. Barcia-Palacios, A. Larrea, Q.A. Parkhurst, C. Martinez, A. Corma, J. Mater. Res. **12**, 1519-29.
20. a. D.-J. Chen, M.J. Mayo, J. Am. Ceram. Soc. **79**, 906-12 (1996). b. Y-S. Her, E. Matijevic, M.C. Chon, J. Mater. Res. **11**, 3121-7. c. R.A. Bley, SM. Kauzlarich, J. Am. Chem. Soc. **118**, 12461-2 (1996).
21. R.R. Bacsa, M. Grätzel, J. Am. Ceram. Soc. **79**, 2185-88 (1996).
22. K.S. Suslick, M. Fang, T. Hyeon, J. Am. Chem. Soc. **118**, 11960-1 (1996).
23. W.R. Moser, B.J. Marshik, J. Kingsley, M. Lemberger, R. Willette, A. Chan, J. Sunstrom, A. Boye, J. Mater. Res. **10**, 2322-34 (1995).

24. M.T. Reetz, W. Helbig, S.A. Quaiser in Active Metals: Preparation, Characterization, Applications, A. Fürstner ed., VCH, Weinheim, 1996, Chapter 7.

25. a. G.L. Messing, S-C. Zhang, G. V. Jayanthi, J. Am. Ceram. Soc. **76**, 2707-26 (1993). b. J.J. Helble, G.A. Moniz, J.R. Morency, U.S. Patent 5,358,695.

26. a. H. Hahn, in Ceramic Powder Science, Ceramic Transactions, Vol. 1., ed. G.L. Messing, E.R. Fuller, H. Hausner, Am. Ceram. Soc., Westerville, OH (1988). b. G. Skandan, H. Hahn, J.C. Parker, Scripta Met. et Mater. **25**, 2389-93 (1991). c. D. Segal, Chemical Synthesis of Advanced Ceramic Materials, Cambridge Univ. Press (1989) p.124. e. H. Gleiter, Prog. Mater. Sci. **33**, 223 (1989).

27. a. G.D. Ulrich, Chem. & Eng. News, August 6, 1984, 22-9. b. S. Vemury, S.E. Pratsinis, L. Kibbey, J. Mater. Res. **12**, 1031-41 (1997). c. S-L, Chung, Y-C. Sheu, M-S. Tsai, J. Am. Ceram. Soc. **75**, 117-23 (1992). d. C-H. Hung, J.L. Katz, J. Mater. Res. **7**, 1861-70 (1992). e. C.-H Hung, P.F. Miquel, J.L. Katz, J. Mater. Res. **7**, 1870-75 (1992). f. P.F. Miquel, C-H. Hung, J.L. Katz, J. Mater. Res. **8**, 2404-13 (1993) i. A.J. Rulison, P.F. Miquel, J.L. Katz, J. Mater. Res. **12**, 3083-89 (1996).

28. a. J.S. Haggerty, R.W. Cannon, J.I. Steinfeld ed. Plenum Press, New York, Chapter 3, (1981). b. M.Cauchetier, N. Herlin, E. Musset, M. Luce, H. Roulet, G. Dufour, A. Gheorghiu, C. Senemaud, Adv. Sci. Tech. **3B** 1319-26 (1995).

29. C-T. Wang, L-S. Lin, S-J. Yang, J. Am Ceram. Soc. **75**, 2240-43 (1992).

30. C.R. Bickmore, K.F. Waldner, R. Baranwal, T. Hinklin, D.R. Treadwell, R.M. Laine, Europ. Ceram. Soc. **18**, 287-97 (1998).

31. Professor H. Hahn, Technische Hochshule Darmstadt, FB Materialwissenschaft, FG Disperse Festoff, personal communication.

32. T. Hinklin, B. Toury, C. Gervais, F. Babonneau, J.J. Gislason, R.W. Morton, R.M. Laine Chem. Mater.**16**, 21-30 (2004).

33. S. Kim, J.J. Gislason, R.W. Morton, X. Pan, H. Sun, R.M. Laine, "Liquid-Feed Flame Spray Pyrolysis of Nanopowders in the Alumina-Titania System, Chem. Mater. In press.

34. J. Marchal, T. Hinklin, R. Baranwal, T. Johns, R.M. Laine, Chem. Mater. **16**, 822-31 (2004).

SIZE AND MORPHOLOGY CONTROL OF CERIUM-TITANIUM OXIDE NANOPARTICLES THROUGH HYDROTHERMAL SYNTHESIS

Jin Lu
Center for Advanced Materials Processing
Clarkson University
Potstdam, NY 13699-5707

J.Davis, P.Seaman, Y.S. Her and X. Feng (*corresponding author: fengx@ferro.com)
Posnick Center of Innovative Technology
Ferro Corporation
7500 E. Pleasant Valley Rd.
Independence, OH 44131

Z.L.Wang
School of Materials Science and Engineering
Georgia Institute of Technology
Atlanta, GA 30332

ABSTRACT

Cerium-titanium oxide nanoparticles were synthesized by a hydrothermal treatment. The particle size, morphology and crystal structure were characterized by laser scattering analyzer, transmission electron microscopy, and X-ray diffractometry (XRD) analyses. The control of secondary and primary particle size as well as particle shape is investigated. The effects of titanium and cerium concentrations, amount and types of bases, reaction temperature and duration on particle size and shape are presented and discussed.

INTRODUCTION

Ceria (CeO_2), either in its pure form or doped with alien cations, has attracted considerable interest because of its wide range of applications including gas sensors, electrode materials for solid oxide fuel cells, oxygen pumps, amperometric oxygen monitors, catalytic supports for automobile exhaust system, and abrasives for chemical mechanical planarizations.[1]

Ceria fine particles can be produced by thermal decomposition,[2] flash combustion,[3] sol-gel,[4] precipitation[5] and hydrothermal methods.[6] In the case of sol-gel and precipitation, calcinations of gels or precursors at relatively high temperature are usually needed to produce cerium oxide, and additional treatment such as grinding and classification is necessary to obtain ultra-fine particles with narrow size distribution.[7]

Hydrothermal synthesis, the method of precipitation from solution under hydrothermal conditions, is attractive for the direct synthesis of crystalline ceramic particles including ceria (doped or undoped) at relatively low temperatures, providing particles with excellent homogeneity and particle uniformity.[8]

For its various applications, ceria particles with different characteristics are required. It is known that the unique chemical and physical properties of nanoparticles are determined not only by their large portion of surface atoms, but also by the crystallorgraphic structures of the particle. The former is determined by the size of the particles, and the latter relies on the particle shape.[9] In present work, the size and shape control for cerium-titanium composite particles (titanium-doped ceria particles) synthesized through hydrothermal treatment is investigated.

EXPERIMENTAL

Particle Synthesis

In a typical preparation, the required amount of cerium (IV) ammonium nitrate $(NH_4)_2[Ce(NO_3)_6]$ (GFS Chemicals, 98.5%) was dissolved in water. The desired amount of titanium isopropoxide (Gelest #AKT872, 99%) was mixed with the above cerium (IV) solution under continuous stirring. The mixed cerium-titanium solution and NH_4OH (28%) were then added to another beaker. The resulting gelatinous pale yellow or brown suspension was heated at >150°C for a few hours in a sealed stainless steel vessel. Thereafter, the vessel was taken out and cooled to room temperature. The reaction product was decanted, washed, filtered and pH adjusted for further measurements and applications.

Particle Characterization

The secondary particle size (aggregate size) was analyzed by laser scattering particle size distribution analyzer (Horiba LA-910). The slurry sample was diluted with deionized water and suspended with Darvan 821A (ammonium polyacrylate, R.T.Vanderbilt Company, 40%) and sonicated for 5 minutes before analysis. The volume-based distribution of particle size was demonstrated by D50, D10 and D90 (diameter on 50%, 10%, and 90%).

The primary particle size and morphology were investigated by high-resolution transmission electron microscopy (HRTEM) using a JEOL 2010, and XRD (Philips PW 3020 Diffratometer) analysis. A holey Cu grid was dipped into the slurry just below the surface to collect some particles, and then the grid was allowed to dry before TEM analysis. Powder X-ray analysis was carried out using Cu-Kα radiation (λ = 1.54056 Å) by backfilling samples into clean holders and collecting the diffractogram over a period from 2θ = 10 to 70 degrees with 0.02 degrees as a step size. The crystallite size was calculated from line broadening of the 220 diffraction peak according to the Scherrer equation, and compared with primary particle size determined from HRTEM measurement (average among 10 to 100 particles). The two results agree well, and the crystal size from XRD was taken as the primary particle size in this work.

The crystallographic and surface structures of ceria particles were determined by HRTEM, performed using JEOL 4000EX transmission electron microscope operated at 400 KV, which is ideally suited for high-resolution structural imaging at a point-to-point Schertzer image resolution better than 0.18nm.

RESULTS AND DISCUSSIONS

In a basic environment, cerium and titanium cations in their starting salts would hydrolyze to form hydroxide or hydrous oxide nuclei, which would convert to a composite oxide via a dissolution-reprecipitation mechanism[10, 11] of crystal growth during hydrothermal treatment. The change of synthesis conditions would affect this nuclei formation and growth mechanism, and as a result, the final particle properties would be changed.

Independent Control of Primary and Secondary Particle Size

A typical nanoparticle generally consists of several primary particles as shown in Figure 1.[12] These primary particles tend to agglomerate into the "secondary" particle, due to their very high surface-to-volume ratio. The strength and size of agglomerates (secondary particles) depend on the surface properties of the nanocrystalline particles, and these properties are sensitively dependent on the powder synthesis conditions.[13] A typical example of titania-doped ceria particle in our study is shown in Figure 2, which demonstrates the weak or loose aggregate of primary particles. In the following discussion, we show how the primary particle and secondary particle sizes of nano-sized cerium oxide can be independently controlled through changes in the starting cation concentrations, base amount (medium pH), reaction duration, and reaction temperature.

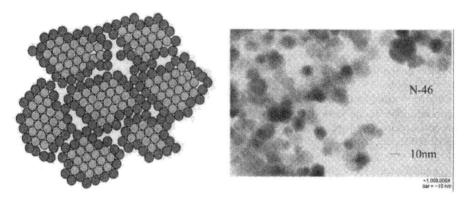

Figure 1. Schematic of "secondary" nanoparticle   Figure 2. TEM image of titanium-doped ceria particle

Starting cation (major metal and dopant metal ions) concentration effect: Cerium concentration has a dominant effect on the secondary particle size, as shown in Figure 3b, and a minor effect on the growth of the primary particle size (crystallite size), as shown in Figure 3a. The primary particle size slightly decreased and then increased with increasing cerium (IV) concentration. In concentrated cerium solutions, because the average diffusion distance for the diffusing solute was short and the concentration gradient was steep, more and more diffusing material passed per unit time through a unit area,[14] favoring more and more crystal growth of ceria particles. When starting with highly concentrated solutions, a large number of particles are formed within the same liquid volume, the mean path of particle collisions is shortened and the high ionic strength also reduces the thickness of the particle double layers, all of which favor the agglomeration of particles. This is why we observed the dominant effects of secondary particle size increase with an increase in cerium concentration as shown in Figure 3b.

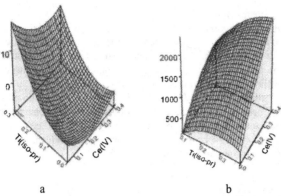

a	b

Figure 3. Computer Interpreted Data: The effect of cation (Ce^{4+} and Ti^{4+}) concentration on primary (a) and secondary (b) particle size of titanium-doped ceria particle. X and Y axes are the concentration values (mole/l), and Z axis is the normalized value of particle size.

Figure 3 also clearly shows that the doping of Ti into the cerium oxide lattice has a dominant effect on the growth of primary particle size, while its effects on the growth of secondary particle size are minimal. With the incorporation of titanium (IV) into the ceria crystal lattice, the mismatch of cation size (ionic radius) and the possible cation valence (Ce^{4+} ↔ Ce^{3+}) between dopant Ti^{4+} and host Ce^{4+} ion, together with the lattice strain and electrostatic interaction by a space charge mechanism,[15] is expected, resulting in the increase of crystal size with more and more Ti^{4+} adding into cerium solution, as shown in Figure 3a. It is also interesting to notice that as the primary particle size is increased dramatically with increasing Ti content in the particles, the secondary particle size gradually decreases. This is probably due to the fact that the growth of the primary particle size is so rapid that the cerium ion in the solution is not enough to feed the crystal growth, and the agglomerated smaller primary particles are also consumed in the growth of the large primary particles.

The above discussion shows clearly that the main control on secondary particle size growth is the host cerium concentration, while the key control for primary particle size is through titanium dopant concentration. Understanding this fundamental growth mechanism enables us to independently control primary and secondary particle sizes.

Effect of base amount (medium pH effect): It is well known that the medium pH has a large effect on particle size growth for many oxides. To study this effect, different amounts of ammonia (in terms of 6x to 12x of the mole amount of cerium ions) were utilized in the synthesis of ceria oxide particles. The excess amounts of ammonium hydroxide were added into the solution to form yellow to brownish geletinous slurries. The effects of the amount of ammonia on primary particle size and secondary particle size are shown in Figure 4.

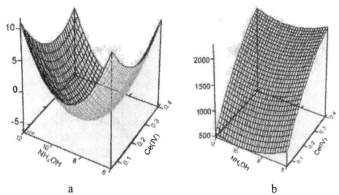

a   b

Figure 4. Computer Interpreted Data: The effect of slurry pH (ammonium amount) on secondary (a) and primary (b) particle size of titanium-doped ceria particle. X and Y axes are the concentration values (mole/l), and Z axis is the normalized value of particle size.

The influence of ammonia concentration on particle size was complicated, and the explanation for the decrease and then increase in crystallite size upon the addition of excessive base has not been clearly established yet. It was suggested that the initial increase in the amount of ammonia (from 6× to 10×) results in a higher concentration of hydroxides. This means that more nuclei are formed under higher pH conditions with the same starting cerium concentration. More nuclei compete for the same amount of cerium supply in the solution, resulting in the low concentration solute in the solution, and thus the rate of crystal growth must be low due to the insufficient supply of the solute by diffusion.[8] Therefore, the final primary particle size decreased. As more ammonia is added (from 10× to 12×), the additional amount of ammonia does not contribute to more hydroxide in the solution due to the limited basicity of ammonia. On the other hand, the large amount of ammonia may enhance the formation of cerium-ammonia complexes on the surface, which retards the dissolution of the primary crystals. If the primary crystal dissolution is hindered while the dissolution-reprecipitation equilibrium established for the growth of the primary crystals still continues, it will result in an increase in primary particle size as the ammonia amount changes from 10× to 12×, as shown in Figure 4a.

An increase in the amount of ammonia decreases secondary particle size, as shown in Figure 4b. This is probably due to the fact that the insufficient supply of cerium in the solution as more ammonia is added also promotes the consumption of the agglomerated small primary crystals to sustain other primary crystallite growth (an effective de-agglomeration). The coating of cerium particles with ammonia through cerium-ammonia complex formation discussed previously also does not favor secondary particle size growth.

Therefore, the addition of different amounts of ammonia provides us with another effective means of adjusting secondary and primary particle sizes.

Hydrothermal temperature and duration effect: It is obvious in Figure 5a that the primary particle size increases proportionately with an increase in hydrothermal treatment temperature (200°C to 300°C). This is probably due to the fact that high temperature provides a high driving force and energy for the growth of the crystals (hydrothermal conversion of cerium hydroxide to ceria is an endothermic process). It is also very interesting to see from Figure 5b that the secondary particle size decreases proportionately with temperature. With the limited amount of cerium in the reaction system, the growth of some of the primary particles has to be at the expenses of some other smaller primary crystals (smaller particles have higher solubility). This process effectively consumes part of the agglomerated primary particles, resulting in smaller secondary particle sizes.

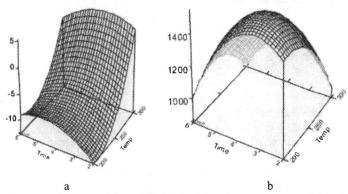

a   b

Figure 5. Computer Interpreted Data: The hydrothermal temperature and duration effect on primary (a) and secondary (b) particle size of titanium-doped ceria particle. X and Y axes are time (hr) and temperature (°C) values, Z axis is the normalized value of particle size.

The reaction duration effects on particle size (both primary and secondary) are much less pronounced in comparison with the temperature effects as shown in Figure 5. The increases in reaction time always increase the primary particle size due to the time needed to build the crystals and the fact that the large crystals built do not re-dissolve as long as the reaction conditions do not change. The secondary particle size shows a more complicated behavior, which needs more understanding in terms of the initial primary and secondary particle size increase and the longer duration decrease in secondary particle size and how this feeds primary size growth.

The temperature of reaction offers us yet another useful control in manipulating primary and secondary particle size.

Particle Shape Control
It is known that nanoparticles usually have specific shape, especially when they are small, because a single crystal nanoparticle has to be enclosed by crystallographic facets that have lower energy. Surface energies associated with different crystallographic planes are usually unique, for face centered cubic structures. For a spherical single-crystalline particle, its surface must contain high-index crystallography planes, which could possibly result in a higher surface energy. Facets formation on the particle surface serves to increase the portion of the low-index planes to minimize the energy of the system. Therefore, small particles of less than 20nm are usually polyhedral.[9]

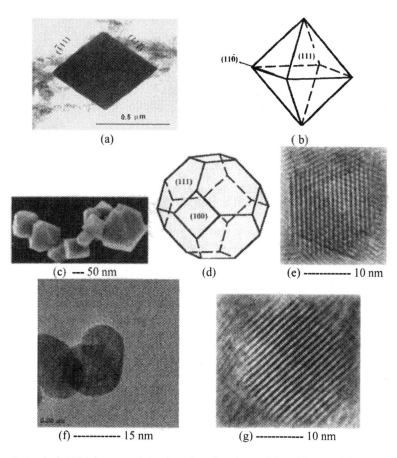

Figure 6. Typical TEM images of titanium-doped ceria particle with urea (a), ammonia (c) and potassium hydroxide (f). Their HRTEM images oriented along {110} (e and g), and structure models (octahedral-b and truncated octahedral-d)

Taking R as the ratio of growth rate in the {100} to that of {111}, Wang et al. had reported[9] that the longest direction in the octahedron (R=1.73) is the {100} diagonal, the longest direction in the cube (R=0.58) is the {111} diagonal, and the longest direction in the cubooctahedron (R=0.87) is the {110} direction. The particles with 0.87<R<1.73 have the {100} and {111} facets, which are named "truncated octahedral". An octahedron has eight {111} facets (Figure 6-a and b), while the truncated octahedron has eight {111} and six {100} facets, and the {100} planes are created by cutting the corners of the octahedron (Figure 6-c, d, e). A truncated octahedron looks like a spherical particle when the particle size is small as shown in f and g of Figure 6.

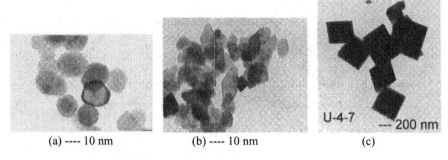

(a) ---- 10 nm   (b) ---- 10 nm   (c)

Figure 7. Typical TEM images of titanium-doped ceria particles made from different bases –
KOH (a), NH₄OH (b) and H₂N-CO-NH₂ (c)

In our nanoparticle synthesis, we can produce particles with different shapes by utilizing different bases (urea, ammonia hydroxide and potassium hydroxide) for the hydrolysis of cerium and titanium salts, as shown in Figure 7. It should be noted that urea produces ammonium and cyanate ions ($H_2N-CO-NH_2 \Leftrightarrow NH_4^+ + OCN^-$) when heated to certain temperature. In neutral and basic solutions, carbonate ions and ammonia are formed($OCN^- + OH^+ + H_2O \rightarrow NH_3 + CO_3^{2-}$). Among the three bases, with the same mole amount of base addition, the case with urea has the lowest concentration of hydroxide ions and therefore the fewest cerium nuclei before hydrothermal treatment, and potassium hydroxide provides the highest hydroxide concentration due to its complete dissociation of hydroxyl group and the most cerium nuclei. The ammonia system is in the middle.

Under the same hydrothermal conditions, the nuclei in a urea reaction system would have the most complete crystal growth, resulting in the formation of the biggest octahedral of any of our reaction systems due to the {111} surface giving the system its lowest energy, as shown in Figures 6a, 6b and 7c. The high concentration of hydroxides in the potassium system favors higher energy surface {100} formation, and the low concentration of cerium due to the high occurrence of cerium nuclei formation also limits the growth of {111} facets.[16] The end result is the formation of a truncated octahedral, as evidenced by the formation of the small round particles (Figures 6f, 6g and 7a). The ammonia system can promote the formation of high energy {100} facets and also has a sufficient cerium concentration to grow {111} facets, resulting in elongated faceted particles as shown in Figures 6c, 6d, 6e and 7b.

CONCLUSIONS

Understanding the particle formation mechanism and properly controlling the experimental conditions allow production of monodispersed CeO_2-TiO_2 nanoparticles through hydrothermal conditions. The primary and secondary particle sizes of these particles can be controlled independently. Precise control is achieved through manipulating the concentrations of starting cerium ions and dopant titanium ions, the amount of base used, the reaction temperature, and the reaction duration. The particle shape can be controlled by using different types of bases such as urea, ammonia, and potassium hydroxide, preferentially promoting low energy surface growth and formation of high-energy facets.

ACKNOWLEGMENTS

The author would like to acknowledge the TEM analysis by Vicky Bryg and SEM analysis by Dave Gnizak of Ferro Corporation. Jin Lu would like to thank Ferro Management for funding her intern work at Ferro Posnick Center of Innovative Technology.

REFERENCES

[1]J.G. Li, T. Ikegami, J.H. Lee and T. Mori, "Characterization and Sintering of Nanocrystalline CeO_2 Powders Synthesized by a Mimic Alkoxide Method," *Acta Materialia*, **49** [3] 419-26 (2001).

[2]P. Janos, M. Petrak, "Preparation of Ceria-based Polishing Powders from Carbonates," *J. Mater. Sci.*, **26** [8] 4062-66 (1991).

[3]M.M.A. Sekar, S.S. Manoharan, K.C. Patil, "Combustion Synthesis of Fine-particle Ceria," *J. Mater. Sci. Lett.*, **9** [10] 1205-06 (1990).

[4]X. Chu, W. Chung, and L.D. Schmidt, "Sintering of Sol-Gel-Prepared Submicrometer Paritcles Studied by Transmission Electron Microscopy," *J. Am. Ceram. Soc.*, **76** [8] 2115-18 (1993).

[5]P.L. Chen, and I.W. Chen, "Reactive Cerium(IV) Oxide Powders by the Homogeneous Precipitation Method," *J. Am. Ceram. Soc.*, **76** [6] 1577-83 (1993).

[6]Y.C. Zhou, and M.N. Rahaman, "Hydrothermal Syntheis and Sintering of Ultrafine CeO_2 Powder," *J. Mater. Res.*, **8** [7] 1680-86 (1993).

[7]Y. Hakuta, S. Onai, H. Terayama, T. Adschiri and K. Arai, "Production of Ultra-Fine Ceria Particles by Hydrothermal Synthesis under Supercritical Conditions," *J. Mater. Sci. Lett.*, **17** [14] 1211-13 (1998).

[8]M. Hirano and M. Inagaki, "Preparation of Monodispersed Cerium(IV) Oxide Particles by Thermal Hydrolysis: influence of the presence of urea and Gd doping on their morphology and growth," *J. Mater. Chem.*, **10** [2] 473-77 (2000).

[9]Z.L.Wang, "Transmission Electron Microscopy of Shape-Controlled Nanocrystals and Their Assemblies," *J. Phys. Chem. B*, **104** [6] 1153-1175 (2000).

[10]G. Stefanic, S. Popovic and S. Music, "Influence of pH on the Hydrothermal Crystallization Kinetics and Crystal Structure of ZrO_2," *Thermochim Acta*, **303** [1] 31-39 (1997).

[11]H. Tomaszewski, H. Weglarz and R. DeGryse, "Crystallization of Yttria under Hydrothermal Conditions," *J. Eur. Ceram. Soc.*, **17** [2-3] 403-06 (1997).

[12]X. Feng, "A Review of Ceramic Nanoparticle Synthesis," *Fine, Ultrafine and Nano Powder 2001 Conference Proceedings,* 75-90.

[13]B. Djuricic and S. Pickering, "Nanostructured Cerium Oxide: Preparation and Properties of Weakly-Agglomerated Powders," *J. Eur. Ceram. Soc.*, **19** [11] 1925-34 (1998).

[14]M. Hirano and E. Kato, "Hydrothermal Synthesis of Cerium(IV) Oxide," *J. Am. Ceram. Soc.*, **79** [3] 777-80 (1996).

[15]S. Lakhwani and M.N. Rahaman, "Hydrothermal Coarsening of CeO_2 Particles," *J. Mater. Res.*, **14** [4] 1455-61 (1999).

[16]Z.L.Wang and X.D. Feng, "Polyhedral Shapes of CeO_2 Nanoparticies," *J. Phys. Chem. B*, **107** [49] 13563-66 (2003).

TRANSPARENT NANOCRYSTALLINE MgO BY LOW TEMPERATURE SPARK PLASMA SINTERING

Rachman Chaim
Department of Materials Engineering
Technion - Israel Institute of Technology
Haifa 32000
Israel

Zhijian Shen, and Mats Nygren
Department of Inorganic Chemistry
BRIIE Center for Inorganic Interfacial
Engineering, Arrhenius Laboratory
Stockholm University, S-106 91
Stockholm, Sweden

ABSTRACT

Nanocrystalline MgO powders were rapidly densified by one-step and two-step spark plasma sintering (SPS) during 5 to 8 minutes at 100 and 150 MPa and temperature range 700 °C to 825 °C. Densification conditions were optimized by SPS at two consecutive temperatures. Fully-dense transparent nanocrystalline MgO with 52 nm grain size was fabricated at 800 °C under 150 MPa for 5 min. In-line transmissions of 40% and 60% were measured for the yellow and red wavelengths compared to MgO single crystal. Comparison of the SPS specimens to those fabricated by hot-pressing indicated the lack of transparency in the latter. This effect is directly related to the fully dense nature of the SPS specimens and the grain and pore size in the nanometer range due to the short processing duration. Rapid densification occurs by particles sliding over each other and leads to the nanometric grain and pore size below the optical wavelength. The light brownish color of the nanocrystalline MgO was related to the oxygen vacancy color centers, originating from the reducing atmosphere of the SPS process.

INTRODUCTION

Most of the high temperature ceramic oxides are optically transparent in their single crystal form. However, the same oxides are translucent or opaque to visible light in their polycrystalline form. This stems from the very fine defects such as porosity and grain boundaries, in the length scale of the optical wavelength, that act as light scattering centres. Full densification of such polycrystalline ceramics for optical transparency requires extended vacuum sintering at very high temperatures (0.6 to $0.8T_m$) depending on the applied pressure.[1] Densification at these conditions occurs by lattice and grain boundary diffusions with very slow kinetics and thus necessitates high temperatures and long durations for full densification of the powder compact. However, surface diffusion with much faster kinetics is always operative at lower temperatures (0.3 to $0.4T_m$). This later mechanism is responsible for sintering and coarsening of the particles but not for their densification. Nevertheless, correct application of this mechanism under applied pressure may result in fully dense transparent ceramics at short durations and relatively low temperatures using spark plasma sintering (SPS).

For this purpose, nanocrystalline ceramic powders are required due to their high specific surface area. These highly active surfaces lower the sintering temperature during pressureless sintering of green ceramic compacts.[2,3] Fundamental defects such as pores may still remain stable in the compact if the pore size is much larger than the particle size.[4,5] This requires preparation of very homogeneous green compacts within which the pore size is of the same order of the nanocrystalline particle size, an impossible task via dry powder technology. Consequently, pressureless sintering of nanocrystalline ceramics at relatively low temperatures,[6] and even for prolonged sintering durations is accompanied by residual porosity. In addition, the slow heating rates (of the order of 2 to 5 °C/min) applied in conventional sintering, provides enough time for particle coarsening up to the sintering temperature. This, in part, results in the loss of the powders' driving force for sintering, hence in partial densification. Recently, translucent MgO ceramics were fabricated by hot-pressing using nano-powders.[7]

On the other hand, a new approach by which the surface diffusion would be maintained as a dominating mechanism during the isothermal densification may be adopted.[8] This necessitates very high heating rates (100 to 500 °C/min) as is used for fast firing[9] or SPS.[10-12] At such high heating rates, the particles are subjected to the transient temperatures for very short durations, thus minimizing the coarsening effects during heating to the consolidation temperature. The particle surfaces may be further activated by dielectric breakdown during field assisted sintering.[13] At this stage, application of external pressure may enhance the densification by sliding of the particles over each other aided by surface diffusion. Theoretically, in these conditions, the original nanostructure grain size may be preserved in the densified compact, provided that the grain boundary and volume diffusion are negligible. In order to implement this theoretical approach, prior knowledge of the densification and grain growth behaviour of the ceramic system is needed and this was acquired from previous hot-pressing experiments.

EXPERIMENTAL PROCEDURE

Pure commercial nanocrystalline MgO (nc-MgO) powder (Nanomaterials Res. Inc., USA) with a specific surface area of 145 m^2/g and calculated spherical diameter of 11 nm was used. The main impurity was 0.01 wt% Fe. The sintering was performed in SPS apparatus, Dr. Sinter 2050 (Sumitomo Coal Mining Co. Ltd., Japan) in vacuum. As-received powder was poured into the graphite die with an inner diameter of 12 mm, and pre-pressed to 150 MPa before heating. The pressure was released and a low pressure of 10 MPa was applied during the heating up procedure. The pressure was increased to 100 MPa within 20 seconds after reaching the final sintering temperature, and a 5 min hold was applied. This procedure was designated as one-step SPS and performed within the temperature range 700 °C to 800 °C. Based on the density results of these series, an alternative two-step SPS was also evaluated in which two consecutive sintering temperatures were used, the conditions of which were summarized in Table I. Generally, the pressures of 20 and 150 MPa were applied for 3 min and 5 min holding times at the lower and the higher sintering temperatures, respectively. This was followed by ending the heating and releasing the pressure. The lower temperature was either 450 °C or 500 °C whereas the upper temperature was between 650 °C and 825 °C. This two-step SPS regime was chosen due to the increase in the densification rate found in hot-pressing experiments above ~ 400°C under the pressure of 150 MPa as will be shown below. The temperature was regulated by a thermocouple that was inserted into the pressing graphite die, at a distance of 2 mm from the sample.

The microstructure of the powder and the dense specimens were characterized using high-resolution scanning electron microscopy (HRSEM, LEO Gemini 982) operated at 4 kV. The specimens were prepared by mechanical polishing using a diamond paste down to 0.25 µm. The surfaces of part of the specimens were chemically etched by immersing in diluted hydrochloric acid solution (10% in water) for one minute. No coating was applied to the specimen surfaces.

HRSEM micrographs were used for image analysis using the program Scion image (NIH). The grain size (diameter) was determined as the largest axis of the grain. At least several hundred grains were counted for determining the grain size distribution. Since the topology of the polished / etched surfaces clearly exposed the true grain morphology, no stereological correction factors were used for the grain size measurements. The median grain size was used for presentation.

Table I: Two-step SPS conditions

Specimen #	1st Step SPS @ 20 MPa for 3 min	2nd Step SPS @ 150 MPa	
	Temp. [°C]	Temp. [°C] *	Time [min]
1	450	825	5
2	450	800	5
3	500	775	5
4	450	775	5
5	450	750	5
6	500	750	3
7	450	700	3

* Final SPS temperature

The density of the specimens was measured by the Archimedes technique using 2-propanol as the immersion liquid.

The optical transmittance measurements were performed on SPS discs of 12 mm in diameter and 1.5 or 1.8 mm in thickness. The parallel surfaces were polished using diamond paste down to 0.25 µm. Real in-line transmission (RIT) was measured using a spectrophotometer (Perkin-Elmer λ-900) where the transmitted light is focused onto the detector.

RESULTS AND DISCUSSION
One-Step SPS

Linear shrinkage of the cold-isostatically pressed nc-MgO green compact (250 MPa, 40% dense) in dilatometer is shown in Fig.1a. The shrinkage began already at 100 °C and reached 3% at 700 °C. However, the main shrinkage (20%) took place in the temperature range 700 °C to 1400 °C. The shrinkage curves under the pressures of 50 and 150 MPa are shown in Fig.1b exhibiting the pressure effect on decrease of the temperature at which the main shrinkage starts. Following the higher shrinkage rate observed above 400 °C at 150 MPa, the hot-pressing experiments were conducted in the temperature range 700 °C to 800 °C and 150 MPa, the pressure was applied already at 550 °C during the heating. The final density - grain size – hot pressing temperature diagram of this nc-MgO is shown in Fig.2.

Generally, the main grain growth during densification starts when the relative density is between 90 to 95% (Fig.2). This is in agreement with the last stage of sintering, where the continuous pores convert to discrete pores that, in turn, are less-efficient in pinning the grain boundaries for grain growth inhibition. Therefore a 'temperature window' exists where rapid densification of the compact is associated with negligible grain growth. Such diagrams are valid for a given powder since both densification and grain growth depend on the powder and green compact characteristics (particle size and porosity distributions). The heating rate in the hot- pressed experiments was 5 °C/min, leading to particle coarsening during heating. Under such hot-pressing conditions, the temperature window may shrink by convergence of the densification and grain growth curves; it may be located at temperatures below 700 °C. However, kinetic studies showed that densification actually was accomplished within the first 60 minutes from the application of the external pressure (curve 'HP' in Fig.3). Therefore, rapid heating of the green compact to temperatures between 700 °C and 800 °C, followed by application of pressure, may result in fully dense nanocrystalline MgO. This approach was tested by SPS of the same nc-MgO powder.

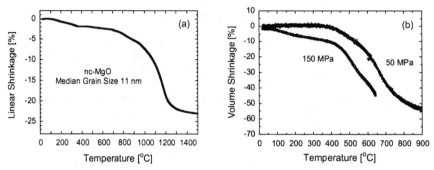

Fig.1. Densification of CIPed nc-MgO compact versus temperature. (a) Linear shrinkage in dilatometer. (b) Volume shrinkage during hot-pressing at different pressures.

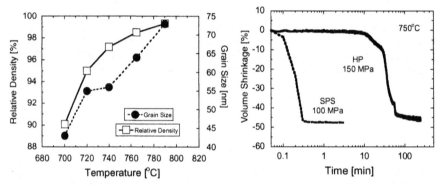

Fig.2. Density-grain size-hot pressing temperature diagram for nc-MgO compacts hot-pressed at 150 MPa for 4 h.

Fig.3. Volume shrinkage versus time during hot-pressing (HP) and SPS of nc-MgO at 750 °C.

The one-step SPS specimens exhibited densities ranging from 91 to 96% by application of 100 MPa pressure for 5 min between 700 °C and 800 °C (open circles in Fig.4a). The present measured densities may represent underestimated values since some graphite impurities (from the graphite die) were captured within the SPS discs. Typical microstructures of the nc-MgO in HRSEM after one-step SPS are shown in Fig.5 and exhibit nanocrystalline grain size together with nanometric pore-size, at the lower SPS temperature (725 °C, Fig.5a). At higher SPS temperature (800 °C, Fig.5b) many submicron size grains were also observed as indicated in the grain size distributions of these specimens (Fig.6). Part of the pores may be due to the chemical etching of the surface. Quantitative image analysis of such micrographs showed that the corresponding median grain sizes range between 30 nm and 70 nm (filled circles in Fig.4a). The grain size distribution of all the specimens was found to be log-normal (see Fig.6), with standard deviation equal to 50% of the median grain size.

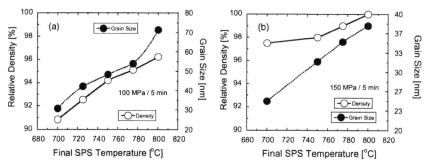

Fig.4. Density-grain size-SPS temperature diagram for nc-MgO compacts densified by (a) One-step and (b) Two-step SPS.

Fig.5. High resolution SEM images of nc-MgO after one-step SPS under 100 MPa for 5 min at (a) 725 °C and (b) 800 °C.

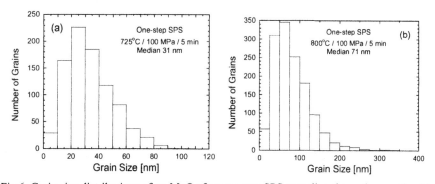

Fig.6. Grain size distributions of nc-MgO after one-step SPS revealing the grain growth with temperature increase where the log-normal distribution character was preserved.

Comparative volume shrinkage during SPS was accomplished within 30 seconds compared to 30 minutes by hot-pressing (Fig.3). Therefore, at the present SPS conditions, significant particle coarsening and grain growth occurred within the very narrow temperature range (between 725 °C and 800 °C) and very short durations, yet without full densification. This prompted us to adopt a two-step SPS process during which higher applied pressure (150 MPa) and additional duration at lower temperature were applied. However, the lower SPS temperature (either 450 °C or 500 °C) was chosen such that the shrinkage rate at this temperature (deduced from hot-pressing conditions; curve 150 MPa in Fig.1b) was still high, similar to that at the final higher SPS temperature.

Fig.7. High resolution SEM images of nc-MgO after two-step SPS under 20 MPa for 3 min at 500 °C followed by 150 MPa for (a) 3 min at 750 °C and (b) 5 min at 775 °C.

Two-Step SPS

The more optimized SPS conditions by which the specimens were densified at two consecutive temperatures resulted in much higher densities as shown in Fig.4b. Systematic increase in the measured density was found at all temperatures, compared to those from the one-step process. The fully dense specimen (# 2 in Table I) was fabricated at 800 °C. The fact that the density at 825 °C was lower than 100% may indicate that optimal SPS temperature exists. In this respect, too high SPS temperature may be associated with extensive particle coarsening during heating, and consequent loss of the effective surface area available for densification via particle sliding. The grain size was also found to monotonically increase with the SPS temperature. Nevertheless, the grain sizes observed (Fig.7) and measured for the two-step SPS (Fig.4b) were smaller than those observed for the one-step SPS (Fig.4a), once again confirming the significance of the particle size on densification. As-expected the grain size exhibited also log-normal distribution (Fig.8)

Optical Behavior

The most important finding was the remarkable optical transparency of the specimens fabricated above 775 °C by single step SPS, and especially those fabricated by the more optimized two-step SPS process, as demonstrated in Fig.9. Sharp and clear images were observed by varying the distance between the (1.5 mm thick) transparent discs and the reflecting paper below them. This was in contrast to the opaque nature of nc-MgO fabricated by hot-pressing.

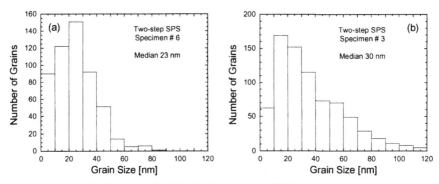

Fig.8. Grain size distributions of nc-MgO after two-step SPS (for the specimens shown in Fig.7) revealing the grain growth with temperature increase where the log-normal distribution was preserved.

Fig.9. Transparent nanocrystalline MgO fabricated by two-step SPS at 20 MPa / 450 °C / 3 min followed by 150 MPa / 825 °C / 5 min. The disc thickness is 1.5 mm. The brownish color is due to the oxygen vacancies present in the disc.

Real in-line transmission versus optical wavelength is shown in Fig.10 for the different specimens, together with MgO single crystal as a reference material. The present results are in agreement with the reported data according to which the measurable transmission region starts at about 300 nm for both polycrystalline and single crystal MgO.[14]

The in-line transmission in the most transparent nanocrystalline specimen (two-step SPS at 825 °C) increased from 40% at 550 nm to 60% at 700 nm, compared to MgO single crystal. Assuming the polished specimen surfaces to be optically smooth, about 16% of the incident light should be reflected from the specimen surfaces using a refractive index of 1.735 for MgO.[15] Therefore, the in-line transmission in Fig.10 represents the net optical transmittance through the specimen that may be affected only by absorption and scattering effects. Absorption effects in the visible range are expected to be negligible due to the ionic bonds in MgO. However, strong effects may be associated with the scattering of light from different types of defects (mainly porosity). In this respect, reflection and refraction of the light beam by the grain boundaries should be negligible, due to the continuity of the refractive index at adjacent grains resulting from the optically isotropic cubic crystal structure of MgO. Consequently, the reduced transmittance is mainly related to Rayleigh scattering by pores,[16] for which the transmitted light intensity is given by:

$$I_T = I_0 \exp(-\alpha_S x) \qquad (1)$$

where I_0 is the initial intensity, α_S is the scattering coefficient, and x is the optical path length.

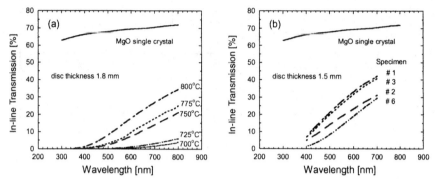

Fig.10. In-line transmission spectra of nc-MgO densified by (a) one-step and (b) two-step SPS at different temperatures, and MgO single crystal as a reference specimen. The decreasing transmission at low wavelengths in the polycrystalline specimens indicates that light is scattered from pores, equal in size to the wavelength.

Under certain boundary conditions, the scattering coefficient due to uniform spherical pores with diameter d_p, that comprise a low volume fraction of the transmitting medium, may be calculated. For a pore size much smaller than the light wavelength λ ($d_p \ll \lambda$) the scattering coefficient may be expressed in a general form as:[17]

$$\alpha_S = A \left(\frac{d_p}{2\lambda} \right)^4 \left(n_{medium}^2 - n_{pore}^2 \right)^2 \qquad (2)$$

where A is constant that relates to the total volume of the pores.

Therefore, scattering strongly depends on the pore size. According to equation (2) the scattering coefficient for a fixed volume of pores increases with the cube of the pore diameter. The scattering coefficient reaches a maximum at pore sizes equal to the light wavelength. However, for pore sizes larger than the incident wavelength, the scattering coefficient decreases in proportion to the inverse of the pore size.[17] This type of dependence leads to very strong scattering when the pore size is within the visible range (400 to 700 nm). The dependence of the in-line transmission on the wavelength follows this expected behavior. Theoretically, a few fraction percents (0.05% to 0.2%) of submicrometer size pores (pore radius of 0.1 to 0.34 μm) with log-normal size distribution (standard deviation = 0.3 to 0.4) should lead to more than a 50% decrease in the transmitted intensity.[18] Comparing the data in Fig.10 to that expected from the theory, and assuming log-normal distribution for the pores (standard deviation = 0.3), porosity levels should be as low as 0.35% in the most transparent specimen, with a maximum pore radius of 100 nm. This is in accordance with the transmittance levels observed in the present work. However, the transparency levels observed in most of the SPS specimens indicate that the present relative densities are strongly underestimated. This may also originate from the fact that the measured densities were normalized by the density of MgO single crystal. This discrepancy in density values may be resolved assuming high volume fraction of disordered grain boundaries having lower density compared to the crystal density.[19]

The light brownish color of the discs was related to the oxygen vacancies formed due to the SPS reducing atmosphere. The formation of large concentrations of oxygen vacancies in MgO subjected to a reducing atmosphere is well known.[20] In addition, when the optical transmittance was low (specimens with single step SPS at 700 °C and 725 °C) the full-scale intensity was changed to lower values (hence with higher sensitivity) to accurately measure the small intensities. Such spectra (not shown here) also contained four very small peaks at the wavelengths 320, 384, 560, and 690 nm. These peaks correspond to photoluminescence at the energies 3.87, 3.23, 2.21, and 1.79 eV, respectively. The peaks at 3.23 and 2.21 eV were directly related to the emission bands from F^+ centers (single-electron trapped oxygen vacancy) and F centers (two-electron trapped oxygen vacancy) in MgO,[20] confirming the oxygen vacancies as the color source of the specimens. These peaks were used to determine the absorption coefficients (34 and 30 cm^{-1} for F^+ and F centers, respectively) from which the concentrations of the F^+ and F centers were estimated as 1.7×10^{17} cm^{-3} and 1.5×10^{17} cm^{-1}, respectively, using equation (1) in ref. 21. Such a defect concentration is expected to affect the color of the present transparent nc-MgO specimens. On the other hand, preliminary investigation of the present specimens by x-ray diffraction indicated no decisive trend of (very small) changes in the lattice parameter, compared to that of stoichiometric MgO. This may originate from the mixed nature of the lattice relaxations around the oxygen vacancies in MgO; small outward relaxation of the Mg^{2+} ions away from the vacancy site for F^+ and F^{2+} centers versus inward relaxation for F centers.[22]

Finally, the transparent nature of the SPS specimens compared to those fabricated by hot-pressing indicated the close to fully-dense nature of the former. This effect is directly related to preservation of the average grain and pore size in the nanometer range below the optical wavelength, due to the short processing duration by SPS at elevated temperatures. Nevertheless, grain growth at densities higher than 95% is an inevitable process due to the loss of solid-gas interfaces.[23] Therefore, refinement of the SPS parameters during low-temperature compaction of nanocrystalline ceramic powders will enable net-shape superfast fabrication of transparent nanocrystalline ceramics as new optical and photonic materials.

ACKNOWLEDGMENTS

Hedva Zipin and Shuki Schechter were acknowledged for the optical measurements and Prof. Wayne Kaplan for the critical review of the manuscript. This work was partially sponsored by the Swedish Research Council through grant 621-2002-4299.

REFERENCES

[1]J. Cheng, D. Agrawal, Y. Zhang, and R. Roy, "Microwave sintering of transparent alumina," *Materials Letters*, **56** [4] 587-92 (2002).

[2]H. Zhu, and R. S. Averback, "Sintering of Nano-Particle Powders: Simulations and Experiments," *Materials and Manufacturing Processes*, **11** [6] 905-23 (1996).

[3]J. R. Groza, and R. J. Dowding, "Nanoparticulate materials densification," *NanoStructured Materials*, **7** [7] 749-68 (1996).

[4]E. B. Slamovich, and F. F. Lange, "Densification of Large Pores: II, Driving Potentials and Kinetics," *Journal of the American Ceramic Society*, **76** [6] 1584-90 (1993).

[5]W. Y. Shih, W-H. Shih, and I. A. Aksay, "Elimination of an isolated pore: Effect of grain size," *Journal of Materials Research*, **10** [4] 1000-1015 (1995).

[6]V. V. Srdic, M. Winterer, and H. Hahn, "Sintering Behavior of Nanocrystalline Zirconia Doped with Alumina Prepared by Chemical Vapor Synthesis," *Journal of the American Ceramic Society*, **83** [8] 1853-60 (2000).

[7]Y. Fang, D. Agrawal, G. Skandan, and M. Jain, "Fabrication of translucent MgO ceramics using nanopowders," *Materials Letters*, **58** [5] 551-54 (2004).

[8]M. Nygren and Z. Shen, "On the preparation of bio-, nano- and structural ceramics and composites by spark plasma sintering," *Solid State Sciences*, **5** [1] 125-31 (2003).

[9]C. Feng, H. Qiu, J. Guo, D. Yan, and W. Schulze, "Fast Firing of Nanoscale ZrO_2 + 2.8 mol% Y_2O_3 Ceramic Powder Synthesized by the Sol-Gel Method," *Journal of Materials Synthesis and Processing*, **3** [1] 25-29 (1995).

[10]Z. Shen, M. Johnsson, Z. Zhao, and M. Nygren, "Spark Plasma Sintering of Alumina," *Journal of the American Ceramic Society*, **85** [8] 1921-27 (2002).

[11]M. Yoshimura, T. Ohji, M. Sando, and K. Niihara, "Rapid rate sintering of nano-grained ZrO_2-based composites using pulse electric current sintering method," *Journal of Materials Science Letters*, **17** [9] 1389-91 (1988).

[12]G-D. Zhan, J. Kuntz, J. Wan, J. Garay, and A. K. Mukherjee, "Alumina-based nanocomposites consolidated by spark plasma sintering," *Scripta Materialia*, **47** [11] 737-41 (2002).

[13]J. R. Groza, M. Garcia, and J. A. Schneider, "Surface effects in field-assisted sintering," *Journal of Materials Research*, **16** [1] 286-92 (2001).

[14]W. D. Kingery, H. K. Bowen, and D. R. Uhlmann, "*Introduction to ceramics*", pp.648-49, 2nd ed., John Wiley & Sons, New York, 1976.

[15]R. J. D. Tilley, "*Colour and optical properties of materials*", p.32, John Wiley & Sons, New York, 2000.

[16]R. Apetz, and M. P. B. van Bruggen, "Transparent Alumina: A Light-Scattering Model," *Journal of the American Ceramic Society*, **86** [3] 480-86 (2003).

[17]H. C. van de Hulst, "*Light scattering by small particles*," Dover, New York, 1981.

[18]J. G. J Peelen, and R. Metselaar, "Light scattering by pores in polycrystalline materials: Transmission properties of alumina," *Journal of Applied Physics*, **45** [1] 216-20 (1974).

[19]P. P. Chattopadhyay, S. K. Pabi, and I. Manna, "On the enhancement of diffusion kinetics in nanocrystalline materials," *Materials Chemistry and Physics*, **68** [1-3] 80-84 (2001).

[20]G. P. Summers, T. M. Wilson, B. T. Jeffries, H. T. Tohver, Y. Chen, and M. M. Abraham, "Luminescence from oxygen vacancies in MgO crystals thermodynamically reduced at high temperatures," *Physical Review B*, **27** [2] 1283-91 (1983).

[21]B. T. Jeffries, R. Gonzalez, Y. Chen, and G. P. Summers, "Luminescence in thermodynamically reduced MgO: The role of hydrogen," *Physical Review B*, **25** [3] 2077-80 (1982).

[22]Q. S. Wang, and N. A. W. Holzwarth, "Electronic structure of vacancy defects in MgO crystals," *Physical Review B*, **41** [5] 3211-25 (1990-I).

[23]N. J. Shaw, "Densification and Coarsening during Solid State Sintering of Ceramics: A Review of the Models, II. Grain Growth," *Powder Metallurgy International*, **21** [5] 31-33 (1989).

CONTROLLED FABRICATION OF NANOMETER-SIZED BUSHES ON INSULATOR SUBSTRATES WITH ASSISTANCE OF ELECTRON BEAM IRRADIATION

Minghui Song[*], Kazutaka Mitsuishi and Kazuo Furuya
High Voltage Electron Miroscopy Station
National Institute for Materials Science
3-13 Sakura, Tsukuba, Ibaraki, 3050003
Japan

ABSTRACT

A new type of nanostructure, bush-like structure, is fabricated with an electron beam irradiation assisted decomposition of organic metal gas. The branches of the nanobush are thinner than 3 nm. The position and size of the nanobush can be controlled by electron beam. Electric charge-up effect is proposed to be a mechanism for the growth of the nanobush structure. Due to a very large ratio of surface area over that of the substrate and controllability in structure, composition, and position, and flexibility in fabrication conditions, the nanobush and its fabrication method are anticipated to be applied in varies fields.

INTRODUCTIOIN

Controlling of structure, composition of materials in nanometer scale and giving them definite properties are methods and also aims of nanoscience and nanotechnology. In this point of view, Nanostructures, such as carbon nanowires, carbon nanotubes, Fullerenes and etc., have obtained much attention these years. Nanostructures are commonly fabricated with chemical vapor decomposition,[1] arc-discharge methods,[2] evaporation,[3] and hydrothermal reaction,[4] etc. However, position controllable fabrication on selected substrates is still a challenge. Here, we report a method to fabricate a new type nanostructure, nanometer-sized bushes (nanobushes) on an insulator substrate by electron-beam-induced deposition (EBID).

To the best of our knowledge, nanobushes containing designed elements grown position controllably on a selected substrate do not exist. The growth of the nanobush is realized by utilizing the feature that an organic gas including metal elements can be decomposed by the irradiation of electron or ion beams. This method has been used to fabricate nanostructures such as nanodots, nanowires, or nano-objects from metal-contained precursors, such as W-, Au-, and Pt-metallorganic chemicals, in a scanning electron microscope (SEM) or a scanning tunneling microscope (STM), or a transmission electron microscope (TEM).[5-12] Compact deposits have been fabricated due to using conductive substrates. However, we used insulators as substrates in the present work. Branches of the nanobush were found to be

---

[*] Corresponding author: Minghui.SONG@nims.go.jp

thinner than 3 nm and from several to over 100 nanometers in length. The size of a bundle of nanobushes itself is controlled by the size of the electron beam used, which is about the same to, or smaller than that of the beam. By controlling the electron beam and supplying gas source, a nanobush with a designed composition can be fabricated at a designed position. Due to a large ratio of the surface area of a nanobush to that of the substrate, the controllability of the composition and position, and the flexibility of fabrication condition, the nanobush and the fabrication method may find applications in nanostructure-making technologies in various fields.

EXPERIMENT

A TEM, JEM2010F (by JEOL Co. Ltd.), equipped with a high bright and small electron beam sized field emission gun, was used for illuminating the substrates by 200 keV electron beam. A carefully designed gas introducing system which has a nozzle with an inner diameter smaller than 0.1 mm and a reservoir of powder gas-sources was used. The system was located inside the TEM. The sublimated gas in the reservoir was slowly leaked from the nozzle located within 2 mm distance from the specimen. Tungsten hexacarbonyl [$W(CO)_6$] powder was used as a precursor material, of which the vapor pressure is approximately 2 Pa at room temperature. The vacuum in the whole column of the TEM was not changed detectably by the gas introduction and was maintained better than 2×10^{-5} Pa. Crystalline and amorphous Al_2O_3 thin films were used as substrates; they make no difference to the fabricated structures. A cold trap filled with liquid nitrogen near the specimen holder was used to eliminate the contamination by the residual gases in the column of the TEM. The fabricated structures were characterized with the JEM-2010F in situ or after detaching the gas source from the specimen holder. All the experiments were performed at room temperature.

RESULTS AND DISCUSSION

A broad electron beam (EB), corresponding to a small current density, is first used to fabricate deposits on Al_2O_3 substrates. Nanometer-sized bush-like structures grew on the substrate within the area irradiated as shown in Fig. 1. This is surprising, since compact deposits have only been fabricated in EBID work till now. The current density in the irradiated area was measured to be about 0.75 A·cm^{-2}. The micrographs were taken at irradiation of 5 and 19 min, respectively (Fig. 1a and b). Contrasts are observed at places on the substrate but apart from the edge in Fig. 1a as indicated by arrows, which are considered to be the nanobushes on the top or back surface of the substrate. These results reveal that the nanobush structure grows in all the area EB irradiated. The branches grew longer under further EB irradiation. At the same time, trunk part is formed. The contrast of the nanobush near the substrate becomes dark in Fig. 1b, indicating that the trunk part becomes more compact than the top part of the nanobush. The branches of the bush are as long as 100 nm.

The thickness of the branches at the top of the bushes is smaller than 3 nm as shown in Fig. 1b.

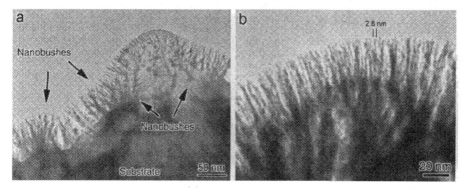

Fig. 1. Nanobushes grown on an Al_2O_3 substrate at current density of 0.75 A·cm^{-2}. (a) The nanobush at 5 minutes of electron beam irradiation. (b) The top part of the nanobush at 19 minutes of electron beam irradiation.

One feature of the nanobush is that the branches grow radially from the bottom of a nanobush to the top. Another is that the nanobush processes a very large ratio of the total surface to that of the substrate where the nanobush grows.

It is naturally found that the size of an area where nanobushes grow is constrained by that of the electron beam used. Fig. 2 gives examples of the area restricted growths of the nanobush. The current density was 7.4 A·cm^{-2} in Fig. 2a and b. Fig. 2a is taken at an irradiation time for 90 seconds showing the diameter of the electron beam of about 150 nm. Nanobushes grew at the edge of the specimen inside the area of the irradiation. After the irradiation time of 150 seconds, Fig. 2b was taken as soon as the electron beam was defocused. The size of the electron beam was not changed from the beginning (at 0 second) to the end (at 150 second) of the irradiation. The area with dark contrast in Fig. 2b corresponds to the area EB irradiated, which is considered to be a compact deposit layer formed on the surface of the substrate. Fig. 2c shows nanobushes grown under irradiation of a more converged electron beam with a diameter of about 50 nm for 5 seconds at each spot. The current density here was estimated as 66.8 A·cm^{-2}. The size of the nanobush grown in the irradiated area becomes small as about 20 nm, but the thickness of branches of the bush at the tip is about the same as that of bushes grown with at small current density.

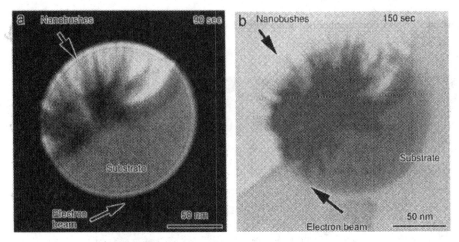

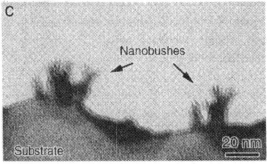

Fig. 2. Nanobushes grown on Al_2O_3 substrates. (a) Showing the size of the used electron beam and the grown nanobush at 90 second of electron beam irradiation. (b) Showing the size of the nanobush grown for 150 seconds. (c) Nanobushes grown under irradiation of a more converged electron beam with a diameter of about 50 nm for 5 seconds at each spot.

It is obvious by comparing Fig. 1 and Fig. 2 that morphology of a nanobush depends on the irradiation condition. In a small current density, the nanobush grows everywhere within the area EB irradiated. While in a definite large current density, the nanobush tends to grow only on a part of the surface within the area EB irradiated. These results indicate that the growth of a nanobush depends on the surface topography of substrate. The nanobush usually grows at the edge of a substrate, especially at a place with a convex surface, as seen in Fig. 2, when the current density becomes large enough. This feature implies that the growth of a nanobush relates to an electric charge-up at the surface of an insulator substrate, since there is a stronger charge-up at a convex place than a concave place on surface of a substrate. This mechanism also explains the growth direction of branches. A charge-up at surface of an insulator substrate could build an electric field inside the vacuum near the surface. The

of the substrate. This is the same direction as the branches of the nanobush grow. This mechanism also explains the feature that the electron beam used can constrain the size of a nanobush structure.

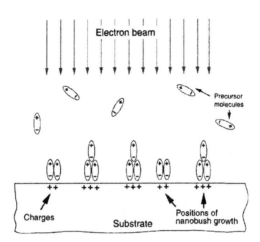

Fig. 3. Schematic graph showing a mechanism for growth of nanobushes on an insulator substrate under electron beam irradiation.

In order to understand the growth process of the nanobush at all the current densities of EB, it is necessary to consider the gas-surface chemistry and the behavior of charges on surface of a substrate. A gas-surface chemistry mechanism of electron- or ion-beam induced decomposition in a beam-assisted chemical etching has been proposed to be composed of two basic processes, namely, an adsorption and a decomposition.[13] It is considered that a molecule of a precursor is first adsorbed on a surface of a sample and then is decomposed into fractions by further irradiation of the energetic beam. It is reasonable to consider that the adsorption and decomposition in EBID are similar to those of a beam-assisted chemical etching process. A molecule of the precursor is decomposed into volatile and non-volatile fractions. The details of the decomposition have been argued to relate to the secondary electrons on surface of a sample produced by incident energetic beam and backscattered electrons in EBID.[5,7] The volatile fraction is pumped away by the vacuum system, while the non-volatile fraction accumulates to form a deposit. A compact deposit is usually formed on surface of a grounded conductive substrate because the compact morphology gives a small surface energy of the deposit. The compact deposit may grow continuously in the area EB irradiated, since the deposit itself is conductive.[5,7] However, for an insulator substrate, charges may be produced on the surface due to secondary electron emission stimulated by incident energetic beams. The charges may accumulate on the surface of the substrate or the deposit, and tend to move to places with convex surface, such as a convex edge, especially sharp tips.

It is reasonable to consider that the charge-up on the surface of an insulator may be not even in a nanometer scale. This unevenness may be resulted from nano-scaled unflatness or atomic steps on surface. Charges may accumulate at some places. Fig. 3 shows this condition of charge distribution schematically. It is supposed that the EB is not very strong at this case, corresponding to the condition in Fig. 1, hence the adsorbed molecules can move around a lot before being decomposed. At a place with denser charges, the built electric field may be stronger than its surrounding place, therefore, movable precursor molecules from both the surface of the substrate and outside the substrate may be attracted to this place, then they are decomposed and form a nonvolatile deposit. After a deposit is formed, charges may move and accumulate at its tip, hence the deposit grows longer under the further EB irradiation. The growth of the nanobush observed in Fig. 1 may be attributed to this mechanism. It can be derived from Fig. 1 that the unevenness of the charge-distribution may be in nanometer scale, and that the density of charges at each places may be not the same, since the nanobush grows separately in nanometers with different speed.

However, when the current density of EB increases to a definite value, precursor molecules adsorbed on surface may be decomposed before moving around. Therefore, a compact deposit layer is formed on the surface within the area EB irradiated, which is considered to correspond to a dark contrast on the substrate in Fig. 2b. This makes the whole area irradiated become conductive. The charges on the surface can move a long way to a convex edge surface or tips. Nanobushes will grow at these places due to a built electric field by the accumulated charges. Branching is also easy to take place at tips of branches due to an accumulation of charges. Therefore, the growth of the nanobush depends on the surface topography of the substrate in submicron scale when the current density is large enough. At the same time, precursor molecules are also adsorbed and decomposed on the surface of the trunk of the bush; hence the bush that is close to the substrate becomes thicker after a longer period of EB irradiation. The above discussion predicts a possibility to fabricate nanometer-sized branched-structures even on a conductive substrate if the substrate is insulated electrically or an electric field is built near its surface.

It was confirmed with energy dispersive X-ray spectroscopy analysis that tungsten, the aimed element, is the main composite of the nanobush. It was also revealed by micro-beam electron beam diffraction that bcc W crystal grains are composed in the nanobush. Further characterization is being carried out and the details will be discussed elsewhere.

There is a high flexibility in fabrication condition of the present nanostructures. Though only $W(CO)_6$ was used to form W containing nanobushes, this technique can be used for other elements such as Mo, Au, Pt, In, and so on, which form sublimating organic powders. It is possible to fabricate nanobush structures containing one or more elements by supplying one or more element sources. Insulator materials other than Al_2O_3 should also be applicable as substrates for fabricating the nanobush structure. An electron beam in intensity from 0.75~ 66.8 $A \cdot cm^{-2}$ has been confirmed to appropriate to fabricate the nanobush structure, which is about the same order as to or smaller than the beam intensities for fabricating nanostructures on conductive substrates by electron beam induced decomposition [5-8, 10,11]. The wide range in current intensity of the electron beam indicates a possibility to use different electron beam sources other than that in a transmission electron microscope for fabricating the nanobush structure. These features make the present method has high potentiality to be used in industry for fabricating nanotree structures of various compositions on various substrates for various applications, such as making nanosized sensors, nanosized catalysts or catalyst supports, and nanosized electronic elements, etc.

CONCLUSION

Nanometer-sized bush-like structures are fabricated on insulator, Al_2O_3, substrates with electron-beam-induced deposition (EBID). The morphology of the structure depends on current density of used electron beam (EB). The structure presents a very large ratio of surface to substrate due to many branches in thickness of about 3 nm and in length up to over 100 nm. The position and size of a bundle of the nanobush can be controlled by controlling those of EB. The growth and the formation of the morphology of the nanobush are explained based on a mechanism involving a charge-up on surface, a uneven distribution of charges, movement of charges to and accumulation of charges at convex surface of substrate or at tips of branched deposits. This process suggests a new method to fabricate new structures with EBID.

REFERENCES

[1]H. F. Zhang, C. M. Wang, E. C. Buck, and L. S. Wang, "Synthesis, characterization, and manipulation of helical SiO_2 nanosprings," *Nanoletters*, **3** [5] 577-580 (2003).

[2]Z. J. Shi, Y. F. Lian, F. H. Liao, X. H. Zhou, Z. N. Gu, Y. Zhang, S. Iijima, H. D. Li, K. T. Yue, and S. L. Zhang, "Large scale synthesis of single-wall carbon nanotubes by arc-discharge method," *J. Phys. Chem. of Solids*, **61** [7] 1031-1036 (2000).

[3]Z. R. Dai, Z. W. Pan, and Z. L. Wang, "Novel nanostructures of functional oxides synthesized by thermal evaporation," *Advanced Functional Mater.*, **13** [1] 9-24 (2003).

[4]Y. Jin, K. B. Tang, C. H. An, and L. Y. Huang, "Hydrothermal synthesis and characterization of $AgInSe_2$ nanorods," *J. Cryst. Growth*, **253** [1-4] 429-434 (2003).

[5]H. W. P. Koops, R. Weiel, D. P. Kern, and T. H. Baum, "High-resolution electron-beam induced deposition," *J. Vac. Sci. Tech. B*, **6** [1] 477-481 (1988).

[6]H. W. P. Koops, J. Kretz, M. Rudolph, M. Weber, G. Dahm, and K. L. Lee,

"Characterization and application of materials grown by electron-beam-induced deposition," *Jpn. J. Appl. Phys. Part 1*, **33** [12B] 7099-7107 (1994).

[7]P. C. Hoyle, J. R. A. Cleaver, and H. Ahmed, "Electron beam induced deposition from $W(CO)_6$ at 2 to 20 keV and its applications," *J. Vac. Sci. Tech. B*, **14** [2] 662-673 (1996).

[8]H. Hiroshima, N. Suzuki, N. Ogawa, and M. Komuro, "Conditions for fabrication of highly conductive wires by electron-beam-induced deposition," *Jpn. J. Appl. Phys.*, **38** [12B] 7135-7139 (1999).

[9]K. T. Kohlmann-von Platen, J. Chiebek, M. Weiss, K. Reimer, H. Oertel, and W. H. Brünger, "Resolution limits in electron-beam induced tungsten deposition," *J. Vac. Sci. Tech. B*, **11** [6] 2219-2223 (1993).

[10]S. Matsui, and T. Ichihashi, "In situ observation on electron-beam-induced chemical vapor-deposition by transmission electron microscopy," *Appl. Phys. Lett.*, **53** [10] 842-844 (1988).

[11]K. Mitsuishi, M. Shimojo, M. Han, and K. Furuya, "Electron-beam-induced deposition using a subnanometer-sized probe of high-energy electrons," *App. Phys. Lett.*, **83** [10] 2064-2066 (2003).

[12]T. E. Allen, R. R. Kunz, and T. M. Mayer, "Monte-Carlo calculation of low-energy electron-emission from surfaces," *J. Vac. Sci. Tech. B*, **6** [6] 2057-2060 (1988).

[13]J. W. Coburn and H. F. Winters, "Ion- and electron-assisted gas-surface chemistry – an important effect in plasma etching," *J. Appl. Phys.*, **50** [5] 3189-3196 (1979).

FORMATION OF NANOCRYSTALLINE ANATASE COATINGS ON COTTON FABRICS AT LOW TEMPERATURE

Walid A. Daoud[*] and John H. Xin
Nanotechnology Centre
Institute of Textiles & Clothing
The Hong Kong Polytechnic University
Hung Hom, Hong Kong

ABSTRACT

Anatase nanocrystalline titania coatings were produced on cotton fabrics from alkoxide solutions using a low temperature sol-gel process under ambient pressure. Titania coatings of the anatase form were obtained via a classical hydrolysis and condensation reaction of titanium isopropoxide that was followed by a hydrothermal treatment. Spectroscopic and microscopic characterizations of the titania thin films showed that the anatase form is predominant throughout the film after the hydrothermal treatment and that the size of grains is about 20 nm.

INTRODUCTION

In recent years, crystalline titanium dioxide has received increasing attention due to their interesting properties and potential applications, e.g. photocatalysts,[1-3] photovoltaics,[4] gas sensors,[5] and electrochromic display devices.[6,7] Nanosized TiO_2 particles show high photocatalytic activities because they have a large surface area per unit mass and volume and hence facilitate the diffusion of excited electrons and holes towards the surface before their recombination.[8] Among the crystalline phases of TiO_2, anatase is reported to have the highest activity.[2] Several techniques such as sol-gel[4], spray pyrolysis,[9] chemical vapor deposition[10] can be used to prepare titanium dioxide thin films. Among these preparation techniques, the relatively simple sol-gel method is the most widely used. However, the disadvantage of all these techniques is that a high temperature process is required to produce anatase thin films.

The formation of photocatalytic anatase titania films at low temperatures is important for the fabrication of transparent films on substrates with poor heat resistance, such as plastics and textiles. Both the optical properties and photocatalytic activity of titania strongly depends on the phase and the size of crystallites. Therefore, the characterization of the microstructure of titania thin films is of high significance.

Recently, Matsuda et al.[11,12] have prepared anatase nanocomposites films by treating an organic polymer (polyethylene glycol) (PEG) modified SiO_2-TiO_2 gel films with hot water at temperatures <100 °C. However, according to these authors, formation of anatase nanocrystals using this method could not be observed in the case of pure TiO_2 and is a unique phenomenon to the SiO_2-TiO_2 system. The authors have also emphasized the role of PEG in the evolution and acceleration of anatase nanocrystals and reported that nanocrystals were formed only on the surface of the coating in absence of PEG.

In the present contribution, we prepared nanocrystalline TiO_2 thin film coatings with high dispersity on cotton at 97 °C using a pure TiO_2 system without the addition of PEG. Spectroscopic and microscopic characterizations showed that after the boiling water treatment,

---

[*] Corresponding author, tcdaoud@polyu.edu.hk

the prepared films were predominantly of anatase form and that the size of grains was about 20 nm.

EXPERIMENTAL

The TiO_2 sol was prepared by mixing titanium tetra-iso-propoxide (Aldrich, 97%) with a solution of absolute ethanol (Riedel, 99.8%), water and acetic acid at room temperature. The resultant mixture was heated to 40 °C and vigorously stirred for 1 hour prior to coating.

A 10 × 10 cm knitted cotton substrate was dried at 100 °C for 5 minutes, dipped in the nanosol for 30 seconds and then padded using an automatic padder at a nip pressure of 2.75 kg/cm^2. The padded substrates were then dried at 80 °C for 10 minutes in a preheated oven to drive off ethanol and finally cured at 97 °C for 5 minutes in a preheated curing oven. The coatings were then hydrotheramlly treated in a boiling water bath for 180 minutes.

The structure and morphology of these coatings were investigated using scanning electron microscopy (SEM) (Leica Stereoscan 440) equipped with (Oxford Energy Dispersive X-ray System) operating at 20 kV and atomic force microscopy (AFM) (Digital Instrument nanoscope, Dimension 3100) in tapping mode.

The crystallinity of the coatings were studied by X-ray diffraction spectroscopy (XRD) (Bruker D8 Discover X-ray Diffractometer) operating at 40 kV, Raman spectroscopy using a 514.5 nm laser line from a CW argon laser (Coherent Innva 70) with 250 mW power, a double grating monochromator (Spex 1403) equipped with a cooled photomultiplier tube (PMT, Hamama-Tus R943-2) and high resolution transmission electron microscopy (HRTEM) (JEOL JSM-2010 microscope) operated at 200 keV.

RESULTS AND DISCUSSION

The observation of the titania films by SEM shows that the surface texture appears to have dense and low porosity characteristics and consists of near spherical grains of about 20 nm in diameter (Fig. 1). Such high dispersity of the nanocrystals could only be observed before in an anatase-precipitated silica coating in presence of PEG using a hot water treatment as reported earlier by Matsuda.[12] Figure 1 also shows the formation of surface aggregates with grains greater than 20 nm in diameter.

The AFM micrograph of titania coating reveals the presence of near spherical particles with diameters of about 35-50 nm as shown in Fig. 2. The particle sizes seem to be rather large compared to those obtained from the SEM results given above. Hence, it seems that the particles observed in the AFM micrograph correspond to agglomerates of small crystallites similar to previous observations.[13,14]

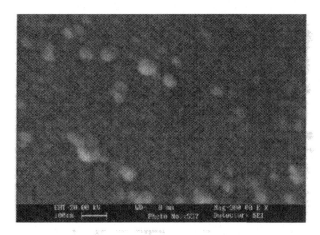

Figure 1. High magnification SEM image of the hydrotheramlly treated titania film coated on cotton.

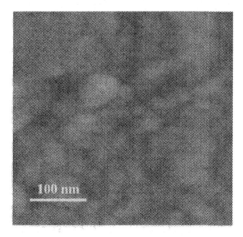

Figure 2. Tapping mode AFM image of the hydrotheramlly treated titania film coated on cotton.

The crystallinity of titania films coated on cotton fabrics was studied by X-ray diffraction (XRD). The bulk of the X-ray signal originated from cotton is the underlying substrate. In the XRD pattern of the as-grown titania films, broad diffraction peaks associated with anatase phase at 25, 37 and 48° were observed in a small magnitude as shown in Fig. 3(b) indicating that the film is predominantly amorphous with a small content of anatase crystallites. This suggests that nucleation of anatase crystallites may have occurred during the sol preparation. Figure 3(c) revealed sharper anatase peaks with greater intensities of the hydrotheramlly treated titania indicating a predominant anatase microstructure of the film.

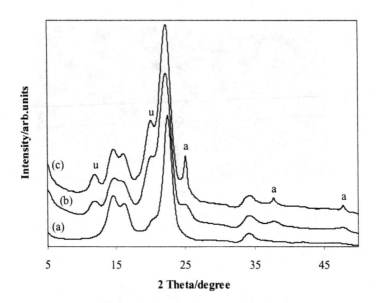

Figure 3. XRD patterns of (a) pure cotton, (b) as-grown titania film coated on cotton and (c) hydrotheramlly treated titania film coated on cotton (a, anatase; u, unknown).

In Fig. 4(b), the Raman spectrum of the as-grown titania films shows peaks associated with the underlying cotton substrates as well as a broad peak at 137 - 161 cm^{-1}. In contrast, the anatase main peak at 150 cm^{-1} was observed after the hydrothermal treatment as shown in Fig. 4(c). It was difficult to observe other anatase-associated peaks as cotton has several peaks in the same region. The Raman spectroscopy indicated changes in the microstructure of the oxide film before and after the hydrothermal treatment are in full agreement with the XRD observations.

Although, HRTEM is a good tool to study the level of dispersion of the nanocrystals throughout the whole coating, it is not practical to perform HRTEM on cotton. Therefore titania films were produced on a silicon wafer by spin coating (1500 rpm for 40 seconds) and heat-treated in a similar manner as explained earlier for coating on cotton. Nevertheless, it must be noted here that the microstructural evolution of the oxide might follow different pathways in each case and that the HRTEM study on silicon substrate is for reference only.

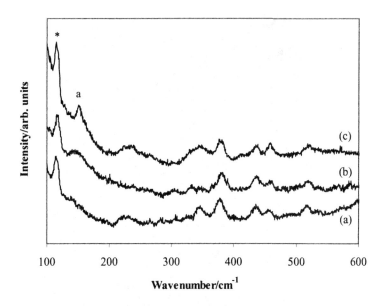

Figure 4. Raman spectra of (a) pure cotton, (b) as-grown titania film coated on cotton and (c) hydrotheramlly treated titania film coated on cotton (a, anatase; *, laser line).

Figure 5(a) shows a HRTEM image and the corresponding selected area electron diffraction (SAED) pattern. The nanocrystals observed in Fig. 5(a) have a lattice fringe of 0.35 nm that corresponds to the 101 lattice plane. Measurements of lattice spacing from the corresponding SAED pattern shown on the top left corner of Fig. 5(a) indicated that the nanocrystals were anatase. Figure 5(b) shows a cross-section HRTEM image of titania film, wherein the dispersion of nanocrystals is quite uniform throughout the coating.

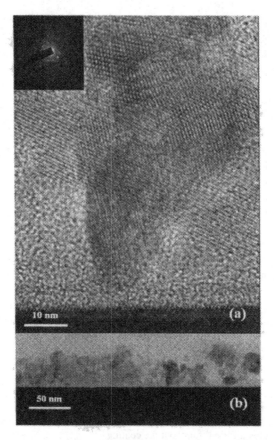

Fig. 5 (a) A high magnification HRTEM cross section image of hydrotheramlly treated titania film coated on a silicon wafer and the corresponding SAED pattern located on the top left corner and (b) A low magnification HRTEM cross section image of titania film coated on a silicon wafer showing the film thickness and grains dispersion. The silicon substrate is located at the bottom of the images.

CONCLUSIONS

Nanocrystalline titania films were produced on cotton fabrics by a low temperature sol-gel process. SEM images revealed a uniform and grainy structure of the films with grains size of about 20 nm. Anatase associated peaks were observed by XRD and Raman spectroscopy. HRTEM images and SAED patterns demonstrated a uniform dispersion of the nanocrystals throughout the titania films. Titania coating on textiles such as cotton is a potential process in the textile industry for use as a self-cleaning antibacterial photocatalyst for the decomposition of organic dirt, environmental pollutants and harmful microorganisms.

REFERENCES

[1]P. Wauthoz, M. Ruwet, T. Machej, and P. Grange, "Influence of the Preparation Method on the Vanadia/Titania/Silica Catalysts in Selective Catalytic Reduction of Nitric Oxide with Ammonia" *Appl. Catal.*, **69** [1] 149-67 (1991).

[2]K. Kato, A. Tsuzuki, H. Taoda, Y. Torii, T. Kato, and Y. Butsugan, "Crystal Structures of TiO_2 Thin Coatings Prepared from the Alkoxide Solution via the Dip-Coating Technique Affecting the Photocatalytic Decomposition of Aqueous Acetic Acid," *J. Mater. Sci.*, **29** [22] 5911-5915 (1994).

[3]P. A. Christensen, T. P. Curtis, T. A. Egerton, S. A. M. Kosa, and J. R. Tinlin, "Photoelectrocatalytic and Photocatalytic Disinfection of E. coli Suspensions by Titanium Dioxide," *Applied Catalysis, B: Environmental*, **41** [4] 371-386 (2003).

[4]B. O'Regan and M. Gratzel, "A Low-cost, High-efficiency Solar Cell based on Dye-Sensitized Colloidal Titanium Dioxide Films" *Nature (London)*, **353** [6346], pp. 737-740 (1991).

[5]P. I. Gouma, M. J. Mills, and K. H. Sandhage, "Fabrication of Free-Standing Titania-Based Gas Sensors by the Oxidation of Metallic Titanium Foils," *J. Am. Ceram. Soc.*, **83** [4] 1007-09 (2000).

[6]U. Bach, D. Corr, D. Lupo, F. Pichot, and M. Ryan, "Nanomaterials-Based Electrochromics for Paper-Quality Displays," *Adv. Mater. (Weinheim, Ger.)*, **14** [11] 845-848 (2002).

[7]P. Bonhote, E. Gogniat, F. Campus, L. Walder, and M. Gratzel, "Nanocrystalline Electrochromic Displays," *Displays*, **20** [3] 137-144 (1999).

[8]M. Anpo, T, Shima, S. Kodma, and Y. Kubokawa, "Photocatalytic Hydrogenation of Propyne with Water on Small-Particle TiO_2: Size Quantization Effects and Reaction Intermediates," *J. Phys. Chem.*, **91** [13] 4305-4310 (1987).

[9]C. R. Bickmore, K. F. Waldner, R. Baranwal, T. Hinklin, D. R. Treadwell, and R. M. Laine, "Ultrafine Titania by Flame Spray Pyrolysis of a Titanatrane Complex," *J. Eur. Ceram. Soc.*, **18** [4] 287-97 (1998).

[10]A. Sandell, M. P. Andersson, M. K.-J. Johansson, P. G. Karlsson; Y. Alfredsson, J. Schnadt, and H. Siegbahn, P. Uvdal, "Metalorganic Chemical Vapor Deposition of Anatase Titanium Dioxide on Si: Modifying the Interface by Pre-Oxidation," *Surface Science*, **530** [1-2] 63-70 (2003).

[11]A. Matsuda, Y. Kotani, T. Kogure, M. Tatsumisago, and T. Minami, "Transparent Anatase Nanocomposite Films by the Sol-Gel Process at Low Temperatures," *J. Am. Ceram. Soc.*, **83** [1] 229-231 (2000).

[12]A. Matsuda, T Matoda, T. Kogure, K. Tadanaga, T. Minami, and M. Tatsumisago, "Formation of Anatase Nanocrystals-Precipitated Silica Coatings on Plastic Substrates by the Sol-Gel Process with Hot Water Treatment," *J. Sol-Gel Sci. Technol.*, **27**, 61-69 (2003).

[13]H. Imai, H. Morimoto, A. Tominaga and H. Hirashima, "Structural Changes in Sol-Gel Derived SiO_2 and TiO_2 Films by Exposure to Water Vapor," *J. Sol-Gel Sci. Technol.*, **10**, 45-54 (1997).

[14]Y. Djaoued, S. Badilescu, P. V. Ashrit, D. Bersani, P. P. Lottici, and R. Bruning, "Low Temperature Sol-Gel Preparation of Nanocrystalline TiO_2 Thin Films," *J. Sol-Gel Sci. Technol.*, **24**, 247-254 (2002).

EFFECT OF ZnO DOPING IN PZT NANOPOWDERS

Ashis Banerjee, Amit Bandyopadhyay and Susmita Bose
School of Mechanical and Materials Engineering
Washington State University
Pullman, WA 99164-2920
E-mail: sbose@wsu.edu

ABSTRACT

Effect of dopants in lead zirconate titanate (PZT) ceramics has been studied by researchers to reduce lead loss and to improve sintering kinetics. Our research is focused on synthesis of ZnO doped PZT nano-powder to study the effect of ZnO dopant on nanopowder, sintering kinetics and properties of PZT system. ZnO doped PZT nanopowder was synthesized using citrate nitrate auto combustion method. Lead nitrate, zirconyl nitrate hydrate and titania are used as starting materials. Zinc nitrate is used as a dopant source in the PZT system. Citric acid is used as a chelating agent as well as source of fuel for the exothermic reaction. Oxidation of citric acid forms CO_2 and H_2O that generates heat and increases the temperature of the system. XRD data shows that the ZnO addition in the range of 0.1 to 5 wt% does not cause any phase transformation. Sintering studies showed that 93% theoretical density could be achieved for doped powders at 900 °C for 1 h. SEM micrograph indicates the formation of a glassy phase at lower temperature in these systems. Room temperature dielectric constant value of 1250, d_{33} of 155 pC/N and k_t of 0.34 are measured with 900 °C sintered compacts.

INTRODUCTION

Lead zirconate titanate (PZT) is one of the most widely used piezoelectric materials due to their excellent piezoelectric properties. PZT based materials can be found in different applications like ultrasonic transducers, hydrophones, speakers, fish finders, sensors, actuators, electrical resonators and wave filters. PZT is a solid solution of ferroelectric $PbTiO_3$ (T_c = 490 °C) and antiferroelectric $PbZrO_3$ (T_c = 230 °C). PZT has a perovskite structure, and below its curie temperature (T_c = 350 °C), it is non-centrosymmetric. The best piezoelectric properties can be obtained in PZT near the morphotropic phase boundary where Zr/Ti ratio is about 1:1.[1-2] The final sintered density, microstructure, doping and composition can influence the properties of the PZT based ceramics. Sintering of PZT at high temperature cause surface oxidation of Pb and forms PbO which creates a lead loss problem at the surface and degrade piezoelectric properties.

Reducing sintering temperature of PZT is one approach to minimize lead loss problem, if excellent piezoelectric properties can be obtained. The most common methods to reduce sintering temperature are use of ultra fine powder[3] or dopants[4-5] to increase densification kinetics. Sintering kinetics is inversely proportional to particle size because of high surface free energy.[6] Additives can also enhance densification by formation of defect structures or low melting glassy phase. Fernandez et al.[7] reported Fe^{3+} doped PZT, a bulk density of 97% of the theoretical density was achieved at 1150 °C. The dielectric constant was 780. Cheng et al.[8] reported Li_2CO_3, Na_2CO_3, B_2O_3, Bi_2O_3 and V_2O_5 as additives to achieve low temperature sintering of Mn and Nb doped PZT. Bulk density was 96% theoretical at 1070 °C. Dielectric constants were in the range of 900 to 1000. Wang et al.[9] reported densification behavior of PZT using a mixture of $LiBiO_2$ and CuO. At 880 °C about 98% densification was achieved, and the dielectric constant were between 1000-1300. Corker et al.[10] investigated the effect of Cu_2O and PbO on the sintering behavior of PZT. At 800 °C about 90% densification was achieved with dielectric constants in the range of 785 - 1010.

We have researched with pure and ZnO doped PZT nanopowder synthesis and characterization. Sintering studies were conducted with these powders at different temperatures. Microstructural analysis and electrical property measurements were conducted on dense samples.

EXPERIMENTAL

PZT nanopowders having 52/48 Zr/Ti composition were prepared by citrate nitrate auto-combustion method. Raw materials used for the preparation of PZT ($Pb(Zr_{0.52}Ti_{0.48})O_3$) were lead nitrate (Sigma, MO), zirconyl nitrate hydrate (Aldrich, WI), titania (Dupont, DE) powder and citric acid (J. T. Baker, NJ). To make a 0.5 M lead nitrate ($Pb(NO_3)_2$) solution, required amount of $Pb(NO_3)_2$ was dissolved in 250 mL distilled water. Then 0.25 M Zr^{4+} ion solution was prepared by dissolving the required amount of zirconyl nitrate hydrate in the 25 mL distilled water with the addition of 225 mL concentrated (15.8 N) nitric acid (HNO_3). This solution was heated to 80-100 °C with continuous stirring until it became clear and homogeneous. To prepare Ti^{4+}-ion solution, first titania (TiO_2) powder was kept for 4 days in hydrofluoric acid (HF). The liquid part was taken out by filtering the mixture. Excess ammonium hydroxide was added to the filtered liquid for complete precipitation. The precipitate was washed with distilled water and then dissolved in concentrated (15.8 N) nitric acid to form a clear solution of Ti^{4+}-ion. To standardize the solution, 10 mL of this solution was taken in a beaker and ammonium hydroxide was added to the solution until complete precipitation. The precipitate was filtered and fired in an alumina crucible at 800 °C for 2 hours. From the calcined product (TiO_2), the amount of Ti^{4+}-ion was calculated. Then by adding distilled water, a 0.5M Ti^{4+}-ion standard stock solution was prepared. PZT precursor solution ($Pb(Zr_{0.52}Ti_{0.48})O_3$) was prepared by mixing required amount of Pb^{2+} solution, Zr^{4+} solution and Ti^{4+} solutions. Total amount of nitrate present in the precursor mixture was calculated from the amount of nitric acid in the mixture, and amount of nitrate that come into the solution from the salts. Then citric acid solution was added to this precursor solution at a citrate to nitrate ratio (C/N) of 0.5 because 0.5 C/N ratio produced minimal agglomeration of PZT. Resulting solution was kept on the hot plate between 80 – 100 °C with continuous stirring. When gelation started, hot plate temperature was increased to 450 °C. After complete drying, the gelatinous mass started burning to form PZT powder. To remove residual carbon, PZT nanopowders were further calcined at 500 °C for 15 min. For ZnO doped nanopowders, required amount of zinc nitrate (J. T. Baker, NJ) was added to the PZT precursor mix and then powder was synthesized using the same citrate nitrate method. The citrate to nitrate ratio was kept at 0.5.

Nanopowders were characterized for their phase purity at room temperature using x-ray diffraction (XRD) analysis with a Philips PW 3040/00 X'pert MPD system using a Co-K$_\alpha$ radiation and a Ni-filter over the 2θ range of 20° to 70° at a step size of 0.02° (2θ) and a count time of 0.5 sec per step. Thermal analysis (DSC/TGA) was done using a STA 409 PC (Netzsch, Germany) system. DSC/TGA experiments were carried out from room temperature to 900 °C in air environment at a heating rate of 5 °C/min. Powder morphology and particle size were evaluated using a transmission electron microscope (JEOL, JEM 120). Particle size distribution was measured using a particle size analyzer (NICOMP 380).

Sintering was done by pressing pellets and then sintered at different temperatures. Microstructural analysis was done using a scanning electron microscope (SEM). Electrical characterization of the highest density samples were done after poling at 100 °C with an applied field of 50 kV/cm. Dielectric constant was measured using a LCZ meter (KEITHLEY 3321, USA), d_{33} was measured using a d_{33} meter (SENSOR 0643 Piezo-d-meter), while k_t was measured using an impedance analyzer (Agilent 4294A, Precision impedance analyzer).

RESULTS AND DISCUSSION

The citrate nitrate auto combustion synthesis is a low temperature synthesis process. This auto combustion process is an oxidation – reduction type exothermic reaction, in which nitrate ions act as an oxidant and carboxyl groups of citric acid act as a reducing agent. The citric acid plays two different roles in this reaction, one is as chelating agent and another is providing the fuel in the auto combustion synthesis. The citric acid holds the metal ions together in the solution by forming a chelated compound and prevents the precipitation of those ions from the solution. Figure 1 shows the complex process of chelation of citric acid with metal ion. The mixture contains large amount of nitrate group. When nitrate starts to decompose it liberates oxygen gas. This oxygen acts as an *in situ* oxidizing agent and helps the carbon to burn which is present in the citric acid. Once carbon starts to burn, CO_2 gas is produced. This exothermic reaction is responsible for increasing the temperature of the mixture.

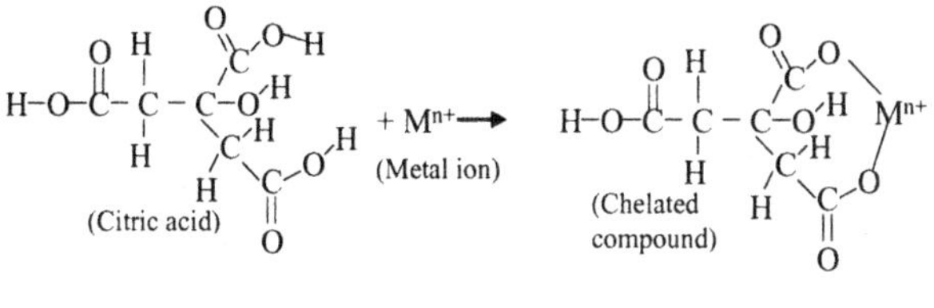

Figure 1. Chelation mechanism of citric acid with metal ion

The increase in temperature helps to burn more carbon in the mixture within a short period of time. Therefore, once the reaction starts, it does not need any further heating from outside. The overall reaction steps are shown below:

$$9Pb^{2+} + 18NO_3^- + C_6H_8O_7 = 9PbO + 18NO_2 + 6CO_2 + 4H_2O$$

$$9Zr^{4+} + 36NO_3^- + 2C_6H_8O_7 = 9ZrO_2 + 36NO_2 + 12CO_2 + 8H_2O$$

$$9Ti^{4+} + 36NO_3^- + 2C_6H_8O_7 = 9TiO_2 + 36NO_2 + 12CO_2 + 8H_2O$$

Figure 2 shows XRD of the pure PZT and ZnO doped PZT powders made by citrate-nitrate synthesis method. Phase pure PZT powder was obtained after calcining at 500 °C for 15 min. Crystallization of PZT occurred before 450 °C on hot plate but calcinations was necessary to form carbon free powder. The XRD plot in Figure 2 indicates that ZnO does not cause any phase transformation in PZT.

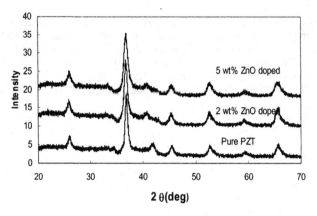

Figure 2. XRD plot of pure and ZnO doped PZT nanopowders.

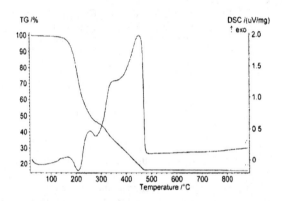

Figure 3. Thermal analysis (DSC/TGA) plot of PZT gel (C/N=0.5).

Figure 3 shows the DSC/TGA plot of the PZT precursor gel that was made with C/N ratio of 0.5. DSC plot shows that there is an endothermic peak at 210 °C. This peak is due to the decomposition of excess citrate ions, which decompose endothermally. Further increase in temperature changes the reaction from endothermic to exothermic and 3 exothermic peaks were obtained at 250 °C, 340 °C and 460 °C. The peak at 460 °C was due to oxidation of carbon. After this peak there is no peak present in the DSC plot. The peak at 250 °C is because of exothermic reaction between citrate and nitrate, and the peak at 340 °C is due to crystallization of PZT. The TGA plot shows there is no weight change after 480 °C, which ensures that no carbon is present above this temperature. Therefore to get carbon free PZT powder, calcinations were done at

500 °C. Figure 4 shows the TEM picture of some freestanding PZT nanoparticles at C/N ratio of 0.5. No clustering or mesoporous type structure was observed in these particles. These particles in the TEM picture are in the size range of 70-90 nm. Increasing the C/N ratio increases the coagulation of particles, which tends to form mesoporous structures.

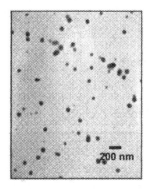

Figure 4. TEM picture of PZT powder prepared at C/N ratio 0.5

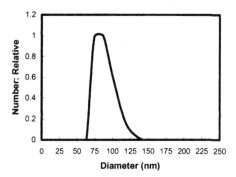

Figure 5. The particle size distribution of PZT powder made by C/N ratio at 0.5

Figure 5 shows the particle size distribution of the PZT nano-powders. The plot shows a narrow particle size distribution in the range of 70 - 110 nm with an average particle size of 90 nm.

Figure 6 shows the effect of ZnO addition on the densification of PZT nanopowders. Figure 6a shows effect of temperature on the densification process. The plot indicates at 900 °C a saturation effect in densification was achieved at 93% theoretical density with 2wt% ZnO addition. The decrease in densification after 900 °C could be speculated because of lead loss and/or trapped air bubble formation in the grain boundary glassy phase. Figure 6b shows the effect of ZnO variation on densification. The plot clearly shows the densification increases with increasing ZnO addition up to 2 wt%, but above 2 wt% no significant change was observed.

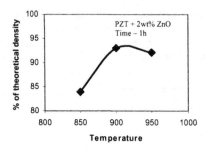 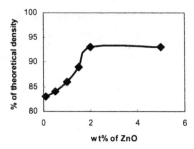

Figure 6. Effect of ZnO on densification. (a) effect of temperature on ZnO doped PZT and (b) effect of ZnO content on the densification of PZT.

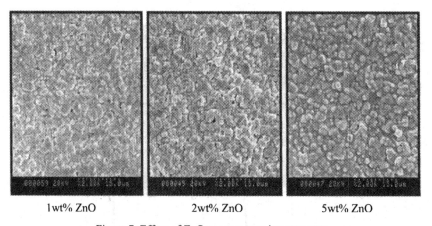

1wt% ZnO   2wt% ZnO   5wt% ZnO

Figure 7. Effect of ZnO content on microstructure

Higher densification at lower temperature with the addition of ZnO can be explained from the microstructures. Figure 7 shows the top surface of the sintered samples. Presence of glassy phase along the grain boundary could be observed due to ZnO addition. The formation of glassy phase increases with increasing amount of ZnO. It is speculated that low melting eutectics are formed due to the addition of ZnO in PZT, which forms a liquid phase during sintering. Energy dispersive spectroscopy (EDS) results showed the presence of ZnO in the glassy phase along the grain boundary, but no ZnO was observed in the PZT grains.

Table I shows the piezoelectric properties of the pure PZT and 2 wt% ZnO doped PZT systems. Pure PZT nanopowder could be sintered at 1200 °C and dielectric constant in the range of 550-600, piezoelectric constant in the range of 150-170 pC/N and coupling coefficient in the range of 0.3-0.34 were measured. ZnO doped PZT nanopowder could be sintered at 900 °C and dielectric constant in the range of 1200-1500, piezoelectric constant in the range of 155-170 pC/N and coupling coefficient in the range of 0.32-0.34 were measured.

Table I. Properties of pure and ZnO doped PZT

Sample	Sintering temperature	Dielectric constant (K)	Piezoelectric constant (d_{33}) pC/N	Coupling coefficient (k_t)
Pure PZT (nanopowder)	1200 °C	550-600	150-170	0.30-0.34
PZT naopowder + 2 wt% ZnO doped	900 °C	1200-1250	155-170	0.32-0.34

CONCLUSIONS

Nanopowders of PZT was synthesized by citrate nitrate autocombustion method. Addition of ZnO along with PZT reduced the sintering temperature to 900 °C compared to 1200 °C of pure PZT nanopowders. 93% theoretical density was obtained after sintering at 900 °C for 1 h. Sintering temperature was reduced due to low melting glassy phase formation because of ZnO addition. Electrical characterization showed that dielectric constant in the range of 1200-1250, piezoelectric constant in the range of 155-170 pC/N and coupling coefficient in the range of 0.32-0.34 for 2 wt% ZnO addition.

ACKNOWLEDGEMENTS

Authors would like to acknowledge financial support from the National Science Foundation (NSF) under the PECASE award to Prof. Susmita Bose (Grant No. CTS-0134476).

REFERENCES

[1]B. Jaffe, R.S. Roth and S. Marzullo, "Piezoelectric Properties of Lead Zirconate – Lead Titanate Solid-Solution Ceramics," *Journal of Applied Physics*, **25**, 809-810 (1954).

[2]K. Kakegawa, J. Mohri, S. Shirasaki and K. Takahasi, "Sluggish Transition Between Tetragonal and Rhombohedral Phases of Pb(Zr,Ti)O$_3$ Prepared by Application of Electric Field," *Journal of the American Ceramic Society*, **65**, 515-519 (1982).

[3]P. Pramanik, R. N. Das, "Structure Property Relations of Chemically Synthesized Nanocrystalline PZT Powders," *Materials Science & Engineering A*, **304-306**, 775-779 (2001).

[4]D. E. Wittmer and R. C. Buchanan, "Low Temperature Densification of Lead Zirconate-Titanate with Vanadium Pentoxide Additive," *Journal of the American Ceramic Society*, **64** [8] 485-490 (1981).

[5]C. Galassi, E. Roncari, C. Capiani and F. Cracium, "Processing and Characterization of High Q$_m$ Ferroelectric Ceramics," *Journal of the European Ceramic Society*, **19**, 1237-1241 (1999).

[6]W. D. Kingery, H. K. Bowen and D. R. Uhlmann, In Introduction to Ceramics, Second Ed; John Wiley & Sons, p 469-477, 1976.

[7]J. F. Fernandez, C. Moure, M. Villegas, P. Duran, M. Kosec and G. Drazic, "Compositional Fluctuations and Properties of Fine Grained Acceptor-Doped PZT Ceramics," *Journal of the European Ceramic Society*, **18**, 1695-1705 (1998).

[8]S. Y. Cheng, S. L. Fu, C. C. Wei and G. M. Ke, "The Properties of Low-Temperature Fired Piezoelectric Ceramics," *Journal of Materials Science*, **21**, 571-576 (1986).

[9]X. X. Wang, K. Murakami, O. Sugiyamaand S. Kaneko, "Piezoelectric Properties, Densification Behavior and Microstructural Evolution of Low Temperature Sintered PZT Ceramics with Sintering Aids," *Journal of the European Ceramic Society*, **21**, 1367-1370 (2001).

[10]D. L. Corker, R. W. Whatmore, E. Ringgard and W. W. Wolny, "Liquid-Phase Sintering of PZT Ceramics," *Journal of the European Ceramic Society*, **20**, 2039-2045 (2000).

PREPARATION AND PROPERTIES OF NANOGRAIN POLYCRYSTALLINE ALUMINA

Marina R. Pascucci and Mark V. Parish
CeraNova Corporation
85 Hayes Memorial Drive Marlborough, MA 01752

William H. Rhodes
Rhodes Consulting
Lexington, MA 02420

ABSTRACT

Polycrystalline alumina (PCA) has great potential for providing performance comparable to or better than single-crystal sapphire, yet offers the opportunity for low-cost powder based manufacturing. High density, large grain-sized PCA is *translucent* due to birefringent scattering of light at grain boundaries – a result of the hexagonal crystal structure of Al_2O_3. CeraNova has demonstrated *transparent* PCA, by processing the material to simultaneously achieve 100% density and sub-micron grain size. The objectionable birefringence is eliminated as the grain size approaches a fraction of the wavelength of light, and the sub-micron grain size leads to high strength and high thermal shock resistance.

Transparent PCA disks have been produced using a powder processing method for green body forming, and a combination of sintering and hot isostatic pressing to achieve 100% density and a fine-grained microstructure (average grain size <0.5µm). CeraNova PCA displays low scatter in the infrared, with high transmittance (>85%) in the 3.0-4.0µm region, comparable to sapphire. Initial mechanical property measurements indicate that CeraNova PCA has toughness ~30% higher than standard alumina and 35-75% higher than sapphire, and hardness ~10% greater than sapphire. For applications where erosion resistance and thermal shock resistance are required, these improvements present significant advantages over sapphire. This fine-grained PCA is a viable sapphire replacement for many applications, including those that require shapes possible by powder processing.

INTRODUCTION

Missile domes are one of the most demanding applications for ceramics. Severe aero-thermal heating occurs as a missile accelerates to its programmed velocity, and the high heating rates require materials with excellent thermal shock resistance. In addition to good mechanical and thermal properties, materials for this application must have high optical transmission over a broad range of wavelengths. Sapphire domes that are currently used have the required combination of properties critical for missile applications. However, shaping sapphire for domes has proven to be very expensive.

Polycrystalline alumina (PCA) has great potential for providing aerothermal performance comparable to or better than sapphire, yet offers the opportunity for low cost powder based manufacturing. Polycrystalline alumina has the same intrinsic properties as sapphire, and the extrinsic property of strength and Weibull modulus are a function of the PCA microstructure. Strength can be made better than sapphire by processing to a fine grain size, and the resulting PCA will have high thermal shock resistance.

High-density, large grain-sized PCA, routinely manufactured commercially for lamp envelopes, is not suitable for dome applications since it is *translucent* due to birefringent scattering of light as it traverses through the many grain boundaries. This intrinsic property results from the hexagonal crystal structure of alumina. Recently, research groups in Germany[1,2] and Japan[3] have demonstrated *transparent* PCA, by simultaneously achieving full density and sub-micron grain size. Objectionable birefringence is eliminated as the grain size approaches a fraction of the wavelength of light.[4] Coincidentally, this same sub-micron grain size requirement works in favor of high strength. Thus, by controlling the grain size for maximum optical properties, strength is also enhanced to levels that can exceed sapphire.

CeraNova's program objectives are to demonstrate fine-grained PCA having 1) superior strength, thermal shock resistance, and abrasion/wear resistance, 2) competitive optical properties, 3) analogous physical properties, and 4) lower cost, as compared to sapphire. Nanograined PCA with excellent transparency and high in-line transmittance has been successfully produced by CeraNova (Figure 1). This material has the potential to meet or exceed the capabilities of sapphire in a variety of demanding applications.

(a)   (b)

Figure 1. CeraNova transparent PCA disks. Two disks polished to 0.8 mm thickness with grain size of ~0.7μm. Focus for (a) is on flag at 1m from disk and for (b) is on SUV at 10m from disk.

PREPARATION OF NANOGRAINED ALUMINA

The overall approach taken by CeraNova to produce transparent ceramic materials involves a low-cost powder processing method with several basic steps: (1) form near net shape by casting a slurry of powder particles, (2) prefire to remove slurry additives and to provide handling strength, (3) sinter to closed porosity with minimal grain growth, (4) HIP to remove porosity and achieve full density with minimal grain growth, (5) lap and polish to optical finish.

Powder Processing/Green Forming

A high-purity α-Al_2O_3 powder (Taimei TM-DAR) with an average particle size of 250 nm (manufacturer's unspecified sedigraph technique) was used for this study. The reported surface area (14.3 m^2/g) would correspond to a smaller particle size. Using the as-supplied powder density of 3.94 g/cm^3, we calculated the particle size to be approximately 100 nm, or about half the average sedigraph particle size. Particle size measurements performed by CeraNova on non-aqueous dispersions using sedimentation (Horiba CAPA 700) confirmed that the majority of the particles (84%) are smaller than 200 nm and that the particle size distribution is quite narrow.

Stable aqueous dispersions of the alumina powder were produced and cast into dense compacts. Slip casting, vacuum filtration and pressure casting were utilized to produce flat, crack-free green disks, up to ~5mm thick. The disks have high and uniform green density (consistently greater than 55 percent) and high green strength. The casting methods used for disks provide a sound basis for casting of non-flat shapes for the dome application.

Disks were dried and pre-fired prior to sintering. The pre-firing temperature was established in a range that was high enough to remove additives and to give sufficient handling strength, but low enough to avoid reducing the powder activity necessary for subsequent sintering.

Sintering

The objective of sintering prior to HIPing was to achieve closed porosity (~95% density) with minimal grain growth. The sintering conditions prior to HIPing are important as they establish the level of porosity, the physical location of the pores (i.e., on the grain boundaries or within the grains), and the grain size. Sintering experiments included the study of several parameters expected to affect the final sintered density.

An initial series of screening sintering runs was conducted in air over a range of temperatures to determine boundaries for achieving closed porosity. Figure 2 shows that sintering at 1225 °C was too low, but samples sintered at higher temperatures reached closed porosity.

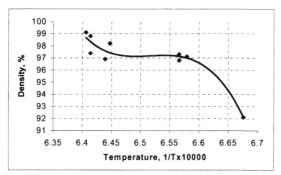

Figure 2. Density as a function of sintering temperature.

Effect of Green Density: When sintering conditions (temperature, time, atmosphere) are held constant, sintered density is a strong function of green density up to a green density of approximately 64%, where it remains constant with further increases in green density. Green densities >56% sintered to closed porosity, samples with green densities >60% were >97% dense after sintering, and several samples in the 61-65% green density range sintered to 99+% density.

Effect of Sintering Aid Composition: Three sintering aids were tested and all samples sintered to closed porosity, with sintered densities >97.5%. A detailed microstructural analysis employing TEM and microprobe to examine the effect of these sintering aids on grain size, porosity, and second phases is in progress.

Effect of Sintering Time: In order to establish baseline conditions for examining other sintering variables, most sintering runs were held at temperature for two hours. However, in many cases, the resulting sintered density exceeded the minimum necessary for closed porosity, and it appeared that the sintering time could be shortened to keep the grain size small while still retaining a closed porosity condition. An experiment to examine the effect of sintering time was conducted at 1275 °C and the results are presented in Table I. All samples reached closed porosity, including those sintered for 0.5h, and the data were consistent from sample to sample.

Table I. Density as a Function of Sintering Time at 1275 °C

Time (h)	Density (%) Sample 1	Density (%) Sample 2
0.5	97.1	97.8
	97.6	
1	98.9	98.9
	99.1	
2	99.5	99.7

Grain Size After Sintering: Grain size after sintering but prior to HIPing was measured on samples which had been sintered under a variety of conditions (Table II). The effects of sintering temperature, sintering time, and sintering aids were examined and all combinations resulted in grain sizes <0.5 μm. It appears that fine-tuning the sintering variables is possible to reduce the grain size to the minimum necessary consistent with achieving closed porosity. Grain sizes after sintering were considerably smaller than the final grain size, but these samples were opaque after sintering and required HIPing to remove the remaining porosity. Representative sintered microstructures (Figure 3) show that remaining porosity appears to be on the grain boundaries, which accounts for the success of HIPing in removing the residual porosity.

Table II. Sintering Parameters, Sintering Aids, and Grain Size prior to HIPing

Sintering Temperature	Time (h)	Density (%)	Sintering Aid	Grain Size (μm)
Low	2	< 95	A	0.35
Intermediate	1	95.6	A	0.31
Intermediate	2	< 95	B	0.24
Intermediate	4	95.2	C	0.46
High	2	95.7	A	0.35
High	2	98.9	D	0.37

Figure 3. SEM photos showing range of grain size for various sintering conditions. Two different sintering temperatures and two different sintering aids. (Marker = 1 μm.)

More recent sintering studies indicate that grain size after sintering does not follow a simple relationship based on sintering temperature and time alone. We hypothesize that grain growth and density are closely related. As more and more grains come in contact, grain growth is free to accelerate. Thus, time at temperature is only part of the story. The parameters controlling grain growth must comprise a density-time-temperature trajectory. This is not a new finding, as early studies by Brook[5] introduced this concept.

Hot Isostatic Pressing

Hot Isostatic Pressing (HIP) was used to achieve full density with a minimum of grain growth. Time, pressure, and atmosphere conditions for all runs were maintained constant, although changes to some of these variables may be examined in the future. In some experiments temperature was examined as a variable.

Initially, HIP temperature and time conditions were kept constant in order to focus on examination of sintering variables. Grain sizes were measured on both the as-HIPed surface and the fracture surface of several samples for comparison with grain sizes prior to HIP. The micrographs in Figure 4 below show that these HIP conditions resulted in significant grain growth. Regardless of the sintering temperature, the grain size after HIP was ~ 0.7 μm.

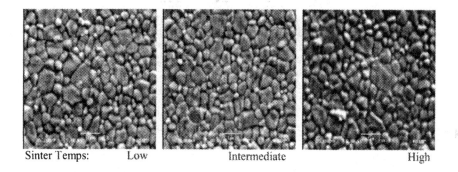

Sinter Temps: Low   Intermediate   High

Figure 4. Grain size after HIPing for samples sintered at several temperatures.
SEM micrographs of HIPed surfaces. Bar = 1 μm.

Additional runs were made at reduced HIP temperatures (time and pressure held constant) to try to reduce final grain sizes. Grain sizes were measured on both the as-HIPed surface and the fracture surface and the data are reported in Table III. In all cases, grain sizes were smaller than for the previous sample group (i.e., <0.7μm), indicating that the final grain size can be reduced by adjusting the HIP conditions. However, contrary to what was expected, the grain sizes at the intermediate HIP temperature are smaller than those observed at the lowest HIP temperature. Possible explanations include inaccuracies in measuring and monitoring the HIP temperature, or perhaps subtle differences in processing parameters which are having a larger than expected effect on the grain size. However, excellent transmission properties have been measured on these samples, which were produced over a range of sintering and HIPing temperatures, and which display a range of resulting grain sizes.

Table III. Relationship between HIPing parameters, sintering aids, and grain size.

HIP Temperature	Time (h)	Sintering Aid	Grain Size (µm)
Low	2	A	0.49
Low	2	B	0.35
Low	2	B	0.36
Intermediate	2	A	0.15
Intermediate	2	B	0.27
High	2	A	0.50
High	2	A	0.67
High	2	B	0.50

CHARACTERIZATION

Optical Properties

Samples were polished for optical characterization at two thicknesses: 0.8 mm and 2.0 mm. Transmittance in the visible (400-2000 nm) was measured with a Perkin Elmer Lambda 900 Spectrophotometer, while transmittance in the infrared (2000-5000 nm) was measured using a Brucker 66IFS FTIR. Figure 5 shows a typical IR transmission spectrum for samples produced early in the program. Two distinct features observed in essentially all spectra are the broad absorption peak centered at 3.3 µm (assumed to be arising from Al-hydroxyl bonds) and another smaller, but well defined absorption peak around 4.2 µm.

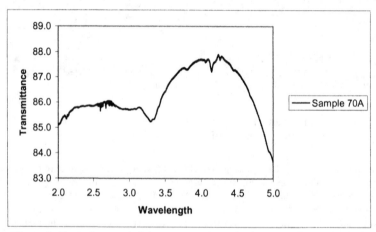

Figure 5. Typical transmission spectra measured using a Brucker 66IFS FTIR.

Table IV compares the measured percent transmittance of CeraNova disks made early in the program (0.8 mm thick) to that of sapphire (3.17 mm thick) and a fine-grained transparent alumina presented by Krell, et al.[6] (0.85 mm thick). Data for the CeraNova material polished to

0.8 mm is comparable to that of Krell, et al. and approaches the values for sapphire, particularly at wavelengths >2.0 μm.

Table IV. Percent transmittance for CeraNova PCA vs. sapphire and other submicron alumina.

Sample Number	Thickness (mm)	645 nm	2.0 μm	2.0 μm	3.0 μm	4.0 μm	5.0 μm
Sapphire	3.17	85.8	86.7	88.8	87.9	87.7	<70
Ref. 6 (est.)	0.85	60		84	85	86	82
61A	0.8	41.9	82.4	84.5	85.8	87.6	83.4
61B	0.8	47.5	82.7	83.8	85.4	87.2	83.1
69B	0.8	55.6	83.1	85.6	85.7	87.5	83.5
69C	0.8	50.9	82.5	85.0	85.7	87.6	83.5
70A	0.8	49.8	82.5	85.1	85.7	87.7	83.7

Effects of Improved Processing, Reduced Grain Size, and OH Removal: Processing improvements implemented throughout this work resulted in more uniform and finer-grained microstructures. In addition, various strategies were adopted to reduce or eliminate absorption bands noted in the FTIR spectra. The results of these improvements can be seen in Table V, where disks from the first stage of the program with average grain size 0.7 μm are compared to improved disks with average grain size <0.5 μm. Also included in the table is data for disk 70A before and after heat treatment for OH removal. Transmission in both the visible and the IR increased with improved processing and a reduction in the grain size.

Table V. Percent transmittance of CeraNova disks with different grain size.

Sample Number	Thickness (mm)	Average Gr. Size	645 nm	2.0 μm	2.0 μm	3.0 μm	4.0 μm	5.0 μm
Sapphire	3.17		85.8	86.7	88.8	87.9	87.7	< 70
69B	0.8	0.7 μm	55.6	83.1	85.6	85.7	87.5	83.5
69C	"	"	50.9	82.5	85.0	85.7	87.6	83.5
70A*	"	"	49.8	82.5	85.1	85.7	87.7	83.7
70A^	"	"	-	-	84.9	86.4	88.0	83.8
81D	0.8	<0.5 μm	59.7	84.6	84.5	86.1	87.6	83.9
81F	"	"	51.1	83.8	85.5	86.5	87.9	84.0
87A	"	"	44.6	83.1	85.7	86.6	88.1	84.0
88A	"	"	61.3	83.9	85.8	86.5	87.8	83.9
88C	"	"	59.7	84.3	87.1	86.7	88.1	84.1

Data for 70A* measured on the as-polished disk; data for 70A^ measured on the same disk after anneal for OH removal.

As noted above, a broad absorption peak centered at 3.3 μm was observed in most samples and was assumed to be arising from Al-hydroxyl bonds. Several samples were heat treated in a controlled atmosphere at a temperature lower than the sintering temperature for a period of approximately one day. An example of the "before and after" IR transmittance spectrum for one of these disks (70A) shows that the annealing treatment resulted in significant reduction of the absorption peak (Fig. 6). Transmission in this region notably increased, whereas transmission in other regions of the spectrum were either unaffected or slightly improved (see Table VI).

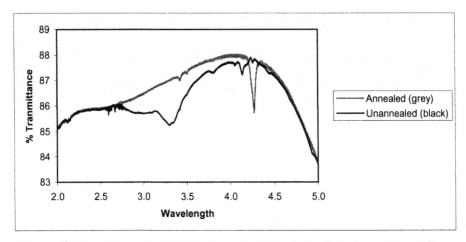

Figure 6. Effect of Annealing on OH Absorption. Black line, before annealing. After annealing (gray line) the broad absorption band centered at 3.3 μm is eliminated.

This phenomenon was studied further using absorbance measurements to determine whether the OH had been completely removed. Two samples, HIPed and polished to 0.8mm and 2mm, were annealed at 1100 °C for various times. After each anneal the absorbance was measured in arbitrary units at 3.3μm and the results are plotted in Figure 7. The data become non-linear after a change in concentration of about 0.5, which is normal according to Crank.[7] The fraction removed is greater for the thinner sample as expected. Calculation yields D ~ 6 × 10^{-8} cm^2/sec, which is about an order of magnitude slower than OH diffusion in Y_2O_3 at the same temperature.[8] The OH is not completely removed after anneals as long as 38 hours. Recent results show that a small, but measurable amount of OH remains even after long-term anneals (~50 hours). Also, transmittance in the IR begins to degrade slightly after long-term anneals, although the reason for this has not yet been determined. Higher annealing temperatures should shorten the time considerably since diffusion is normally exponentially related to temperature. However, this must be done carefully to avoid degrading the overall in-line transmittance.

One of our goals is to produce transparent material that is several millimeters thick, both for the intended dome application and also for fabrication of mechanical test specimens. Numerous transparent disks have been fabricated and polished to 2 mm thickness, and the transmittance results are presented in Table VI. For comparison, the measured transmittance for sapphire 3.17 mm thick is included, along with calculated transmittance for 2 mm thick sapphire. Disks 70–75 were produced early in the program and, while the transmittance values approach those of sapphire, the disks still display evidence of scattering and birefringence, which reduces the transmittance. Disks 81-89 were produced later in the program and show the effects of improved processing. Transmittance for these disks in the 2-5 μm region is significantly improved, and the value at 4 μm is within 2% of (or in some cases equivalent to) sapphire.

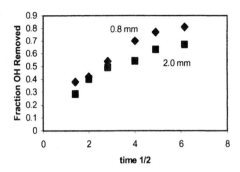

Figure 7. Fraction OH removed versus annealing time, t ½.

Table VI. Transmittance for 2mm samples. Disks 81-89 show the effects of improved processing. Transmittance in the IR region is significantly improved.

Sample Number	Thickness mm	645nm	2.0µm	2.0µm	3.0µm	4.0µm	5.0µm
Sapphire (measured)	3.17	85.8	86.7	88.8	87.9	87.7	< 70
Sapphire (calculated)	2.0				88.0	88.2	
70C	2.0	22.7	74.3	74.5	76.9	81.0	71.9
71A	"	16.2	72.1	73.6	79.4	84.1	74.4
73C	"	34.4	79.0	81.2	82.3	85.6	74.9
75A	"	23.0	76.3	78.7	81.6	85.5	75.0
81C	2.0	24.2	80.0	82.7	84.5	86.9	75.2
86B2	"	42.2	82.3	85.8	86.9	88.6	76.5
86C	"	35.7	81.8	84.6	85.8	87.4	75.4
87B	"	16.0	78.8	82.3	86.4	88.7	76.8
88D	"	33.3	81.0	83.6	86.1	88.4	76.6
89A	"	34.7	80.6	82.5	85.2	87.7	76.0
89C	"	31.8	80.6	83.3	84.8	86.9	75.2

Mechanical Properties

Indentation hardness tests were performed and the data were used to calculate fracture toughness. Figure 8 shows Vickers hardness indents for lighting grade translucent PCA (grain size 25-30 µm) and fine-grained transparent CeraNova PCA. Samples were indented using a 1kg load, and the hardness and toughness values are an average of data from 12 indents. The measured hardness of fine-grained PCA is about 20% greater than that for standard PCA and 10% greater than that of sapphire. Sub-surface cracking in the standard PCA precludes using the indentation technique to calculate toughness for this material. The toughness value calculated for fine grained PCA, $K_{Ic} = 3.5$ MPa m^{½}, is comparable to literature values reported for similar materials and is about 30% higher than the toughness of standard alumina, and 75% higher than that of sapphire.[9] Future work will include strength testing of the fine-grained PCA and an estimate of the potential improvement in thermal shock resistance versus sapphire.

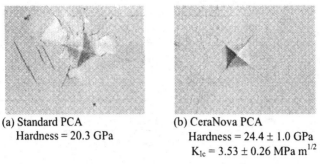

(a) Standard PCA
Hardness = 20.3 GPa

(b) CeraNova PCA
Hardness = 24.4 ± 1.0 GPa
$K_{Ic} = 3.53 \pm 0.26$ MPa m$^{1/2}$

Figure 8. Vickers indentation for "standard" PCA (a) and CeraNova fine-grained PCA (b). Fine-grained PCA shows no sub-surface damage and has higher hardness and fracture toughness.

CONCLUSIONS AND FUTURE WORK

An important focus of future work will be to tailor the grain size to further improve the optical properties. Apetz and van Bruggen[4] generated a model for predicting in-line transmission as a function of grain size and sample thickness. Figure 9 (after ref. 4) shows the excellent agreement between their model and their experimental data for grain size G=0.5 μm and sample thickness d=0.8 mm. Using this model, CeraNova has calculated the theoretical transmission for two additional grain sizes: G=0.7 μm and G=0.2 μm, i.e., the largest grain size in transparent samples made by CeraNova after HIP and the smallest grain size achieved after sintering. According to the model, significant improvements in transmission can be achieved, particularly in the visible, as the final grain size in a fully dense sample is reduced below 0.5 μm.

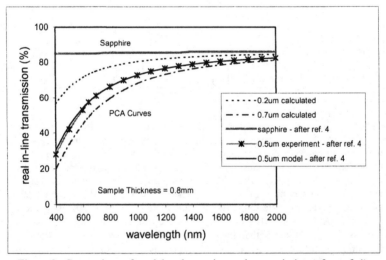

Figure 9. Comparison of model and experimental transmission (after ref. 4).

CeraNova has demonstrated a process for making state-of-the-art fine grained transparent PCA and has achieved the following goals: 1) low scattering and good in-line transmittance in the infrared (3-5 μm); 2) production of transparent, nanograin polycrystalline aluminum oxide samples several millimeters thick using a low cost powder processing method; and 3) optical transmittance within two percent of the transmittance values of sapphire. Initial mechanical property measurements (hardness and indentation toughness) indicate that CeraNova's fine-grained PCA is comparable to similar materials reported in the literature and superior to sapphire. The uniform polycrystalline microstructure, increased hardness and toughness, and potential for near-net shape processing for CeraNova PCA present significant advantages for this material in dome applications.

Nanograin polycrystalline aluminum oxide makes possible a new and exciting low cost missile dome material for many DOD applications in addition to opening commercial applications that have been cost prohibitive. The PCA demonstrated by CeraNova is a viable sapphire replacement for many applications, including those that require shapes possible by powder processing and not possible by the grinding and polishing of sapphire.

ACKNOWLEDGEMENT

This work was partially supported by Navy contracts N68936-03-C-0011 and N68936-03-C-0089.

REFERENCES

[1]A. Krell, "Fracture Origin and Strength in Advanced Pressureless-Sintered Alumina," *J. Am. Ceram. Soc.*, **81** [7] 1900 (1998).

A. Krell and E. Strassburger, "High-Purity Submicron αAl₂O₃ Armor Ceramics Design; Manufacture; and Ballistic Performance," *Am. Ceram. Soc. Bull.*, **80** [7] 100 (2001).

A. Krell, P. Blank, H. Ma, and T. Hutzler, "Transparent Sintered Corundum with High Hardness and Strength," *J. Am. Ceram. Soc.*, **86** [1] 12-18 (2003).

A. Krell, P. Blank, H. Ma, T. Hutzler, and M. Nebelung, "Processing of High-Density Submicrometer Al₂O₃ for New Applications," *J. Am. Ceram. Soc.*, **86** [4] 546-53 (2003).

[2]D. Godlinski, M. Kuntz, and G. Grathwohl, "Transparent Alumina with Submicrometer Grains by Float Packing and Sintering," *J. Am. Ceram. Soc.*, **85** [10] 2449-56 (2002).

[3]K. Hayashi, O. Kobayashi, S. Toyoda, and K. Morinaga, "Transmission Optical Properties of Polycrystalline Alumina with Submicron Grains," *JIM*, **32** [11] 1024 (1991).

[4]R. Apetz and M.P.B. van Bruggen, "Transparent Alumina: A Light Scattering Model," *J. Am. Ceram. Soc.*, **86** [3] 480-86 (2003).

[5]R.J. Brook, "Controlled Grain Growth," in Treatise on Materials Science and Technology, Vol. 9 Ceramic Fabrication Processes, Franklin F.Y. Wang, ed., Academic Press, New York (1976). p. 331.

[6]A. Krell, G. Baur, and C. Dähne, "Transparent Sintered Sub-μm Al₂O₃ with IR Transmittivity Equal to Sapphire," Proceedings SPIE, April 2003, Orlando, FL.

[7]J. Crank, The Mathematics of Diffusion, 2ⁿᵈ: Clarendon Press, Oxford, 1975.

[8]G. C. Wei, "Extrinsic OH⁻ Absorption in Transparent Polycrystalline Lanthana-doped Yttria," *J. Am. Ceram. Soc.*, **71**, C20, (1988).

[9]D. Harris, "Materials for Infrared Windows and Domes, Properties and Performance," SPIE Optical Engineering Press, 1999, p. 117-118.

SYNTHESIS OF Nb and La DOPED TiO$_2$ FOR GAS SENSORS

Young Jin Choi, Ashis Banerjee, Amit Bandyopadhyay and Susmita Bose
School of Mechanical and Materials Engineering
Washington State University
Pullman, WA 99164-2920
E-mail: sbose@wsu.edu

ABSTRACT

Nanostructured TiO$_2$ based materials for gas sensors were synthesized using citrate-nitrate auto combustion method as a function of different dopants. Titanyl nitrate solution was prepared using commercial TiO$_2$ powder, hydrofluoric acid (HF) and concentrated nitric acid (HNO$_3$). The amount of template material was optimized to produce TiO$_2$ nanoparticles. La$_2$O$_3$ and Nb$_2$O$_5$ were used as dopants in TiO$_2$. X-ray diffraction analysis confirmed formation of anatase phase TiO$_2$. The BET specific average surface area analysis showed a decrease in surface area with the addition of dopants and as the dopant concentration increased, surface area decreased. Conductivity of these discs was measured in the temperature range between 300 and 800 °C in the furnace air environment as a function of composition. Particle size analysis showed that the size varied between 100nm and 400nm for pure TiO$_2$ and doped TiO$_2$. Our initial results indicate that dopants chemistry and amount has significant influence on the conductivity for these nano-structured TiO$_2$ based ceramics.

INTRODUCTION

Synthesis and processing of nano-size particles have attracted a great deal of interest because nanoparticles have a unique ability to improve mechanical, electrical, optical and chemical properties compared to conventional micrometer sized polycrystalline materials.[1] In gas sensor applications, sensor sensitivity could be improved using nanocrystalline materials.[2] To obtain optimum and reproducible nanosized materials, we have used modified pechini method, which is an auto combustion process at low temperature.[3] Lessing et al. reported that pure and Ce-doped SnO$_2$ nanoparticles for gas sensor could be prepared by polymeric precursor method based on the pechini process using tin citrate aqueous solution, which was prepared from SnCl$_2$·2H$_2$O.[4] In this process, ignition of the resin initiates at a relatively low temperature that remove the organic material, and leaves the desired compositions in nano-powder form.[5] In the present research we have used the citrate-nitrate auto combustion method to obtain nanosize TiO$_2$ powders.

Seiyama (1962) and Taguchi (1970) first applied the gas detection techniques in gas sensors and since then it has found wide-range of applications in domestic and industrial gas detectors for gas-leak alarm, process control and pollution control.[6-7] For ceramic gas sensors, transition metal oxides are most widely used. A change in their electrical conductivity due to the presence of the sensing gas is typically used for sensing measurements. Among different transition metal oxides, the most widely studied gas sensor materials are SnO$_2$ based. As gas sensitive resistors, these sensors show good sensitivity and selectivity below 250 °C temperature. However, at higher temperatures, these materials show poor sensitivity and low stability.[8] Thus there is a need to find reliable high temperature sensor materials. Akbar et al. reported that TiO$_2$ based gas sensors show good sensitivity and stability in adverse environments such as at 600 °C.[9-10] TiO$_2$ based gas sensor can be used even up to 800 °C, if one can stabilize the desired phase of TiO$_2$.

Nanostructured TiO_2 based gas sensors can have even greater potential to gas sensing properties. Because of the grain size reduction, increases in average specific surface area, which improve gas-diffusion control and result in better gas sensing properties. It was reported that small amounts of dopants can affect gas sensing properties,[12-14] because dopants can be added in TiO_2-based sensor materials to stabilize a phase, or to improve sensing properties.[15-19] In this paper, we report synthesis of nanosized TiO_2 materials using citrate nitrate auto combustion method. We have characterized these materials for their phase, surface area and particle size distribution along with preliminary studies on their high temperature conductivity as a function of dopant composition in furnace air environment.

EXPERIMENTAL
Synthesis of TiO_2 Nano-Powders
All samples were synthesized from commercial TiO_2 powders (Dupont, Santa Fe Springs, CA). During synthesis, La and Nb nitrate salts are used to achieve La_2O_3 and Nb_2O_5 doped TiO_2 nanopowders. Commercial TiO_2 powders were kept for 4 days in HF solution to prepare Ti^{4+} solution. After 4 days, the liquid part was taken out by filtering the mixture. Excess ammonium hydroxide (NH_4OH) was added to the filtrate for complete precipitation at pH 10. The precipitate was washed using distilled water and then dissolved in concentrated nitric acid (HNO_3) to form a clear Ti^{4+} solution. The citric acid solution was then added to this solution with a cirtrate to nitrate ratio (C/N) of 0.5. Aqueous solutions of La^{3+} and Nb^{5+}, using their respective nitrate salts as needed, was added to this solution. The solution was heated on the hot plate and after the gelation was complete, the hot plate temperature was increased for complete drying. After complete drying, the gelatinous mass started to burn. A final calcination was done in alumina crucible at 700 °C for 1hr and 2hr. After calcination, these TiO_2 nano-powders were characterized for their phase stability, BET specific average surface area and particle size distribution.

Disc Preparation
TiO_2 nanopowders were used to process green compacts via uniaxial compression. Powders were first pressed into discs of 0.06cm thickness and 1.27cm in diameters. The discs were densified at 800 °C for 1hr. Since the anatase to rutile phase transformation takes place between 800 and 900 °C in air, 800 °C heat treating temperature was selected to avoid any rutile phase formation. For conductivity measurements on these discs, silver paint was used as electrode.

Characterization of Pure and Doped TiO_2 Nanopowder
TiO_2 nano-powders were characterized for phase analysis using a XRD (Philips PW 3040/00 Xpert MPD) system. Surface area was measured using a five point BET surface area analyzer (Tristar 3000, micromeritics, GA). Scanning electron micrographs (SEM) were taken on a Jeol 6400 electron microscope. Particle size analyzer (Nicomp 380, Santa Barbara CA) was used to measure particle size of the synthesized powder. Thermal analysis was done using NETZSH STA 409 PC to understand the decomposition behavior during TiO_2 powder formation. For conductivity measurement, samples were placed in a tube furnace in furnace air environment. Testing temperature ranged from 400 to 800 °C. Resistance was measured using a multimeter (KEITHLEY 2400 Digital Source Meter).

RESULTS AND DISCUSSION

Figure 1 shows the differential scanning calorimetry (DSC) data for synthesized TiO_2 nanopowders after heat treatment on hot plate at 450 °C. There are two broad peaks in the DSC plot, the first between 630 and 670 °C, and the second between 800 and 850 °C. The first exothermic peak is due to the oxidation and decomposition of the released small organics and inorganics during the synthesis. Excess carbon burns out in this step. The other peak, an endothermic peak, shows that the phase transformation from anatase to rutile takes place between 800 and 850 °C in furnace air environment. We focused our study on the anatase TiO_2, since it has shown excellent sensing properties.[9-11] Sintering below 800°C avoids the transformation of anatase to rutile.

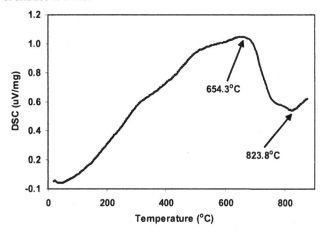

Figure 1. DSC plot for the TiO_2 nano-powders after heat treatment at 450 °C on the hot plate with citrate/ nitrate ratio of 0.5.

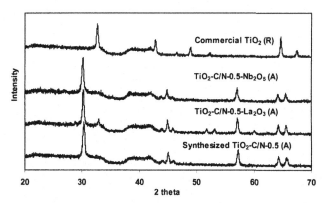

Figure 2. Powder X-ray diffraction patterns for commercial TiO_2 powder, synthesized doped and undoped TiO_2-powders with citrate/nitrate ratio if 0.5 and heat-treated at 700 °C for 1hr. (R-rutile, A-anatase).

X-ray diffraction (XRD) results are shown in Figure 2. Results show that all synthesized TiO$_2$ nanopowders, with and without dopants, primarily contain anatase phase. Commercial TiO$_2$ powder, the starting material, had the rutile phase. It can be seen from the XRD data of commercial TiO$_2$ that HF dissolving can influence phase change from rutile to anatase. It can also be seen that dopant weight%, as high as 15wt%, did not have significant effect on the anatase phase formation for these nano-TiO$_2$ powders.

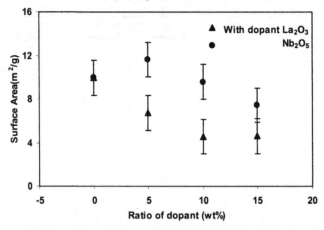

Figure 3. Changes in the specific surface area of the TiO$_2$ powders obtained by different dopants (La$_2$O$_3$ and Nb$_2$O$_5$) with calcination temperature of 700 °C for 1hr.

Figure 3 shows change in BET specific average surface area as a function of dopant concentrations. It can be seen that increasing amount of dopants has some effect on the surface area of these nanopowder. It is known that the sensitivity can be improved by increasing the surface area, which will increase the number of oxygen sites on the surfaces.[12] Thus, a higher surface area is desirable in the powder to make gas sensors which will employ surface conducting mechanism for gas sensing. Nanoparticles have high tendency to agglomeration due to their very high surface energy.[20] Figure 3 shows decrease in BET surface area as La$_2$O$_3$ and Nb$_2$O$_5$ dopant concentration increases in pure TiO$_2$ nanopowders. It is common to introduce dopant such as Nb and La to TiO$_2$ crystals to provide an increase in its conductivity.[21] These dopants may substitute Ti in TiO$_2$ crystal structure or stay at an interstitial site, and modify specific surface area. It means that dopants play an important role in incorporation of impurity.

Particle size analysis shows that for pure and La doped TiO$_2$ samples particle size varies between 150 and 400nm with wide particle size distribution whereas for Nb doped TiO$_2$ samples the particle size varied between 125 and 225nm with a narrow size distribution as shown in Figure 4. Average particle size generally decreases when pure TiO$_2$ was synthesized with dopants. It is also found that the addition of La$_2$O$_3$ and Nb$_2$O$_5$ increases the densification rates, as shown in Figure 5, which is also reported in literature.[13-18]

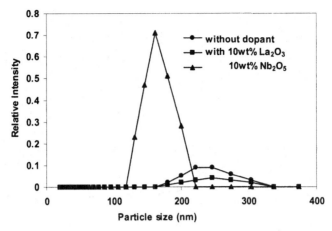

Figure 4. Particle size analysis of TiO₂ powder with and without dopants.

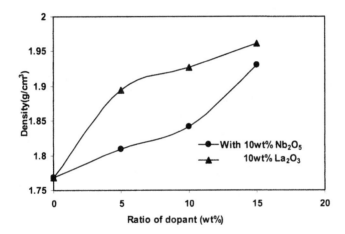

Figure 5. Changes in the density of the TiO₂ powders obtained by different dopants (La₂O₃ and Nb₂O₅) with sintering temperatures of 800 °C for 1hr.

In order to compare the morphology of TiO₂ nano powders, with and without dopants, microstructural analysis was carried out using SEM and shown in Figure 6. Figure 6a shows the microstructure of pure TiO₂ disc sample sintered at 800°C for 1hr. Significant amount of porosities are found in these samples as expected, and particle size was varied between 250 nm and 1 μm. Some of them are also agglomerated. Figure 6b shows Nb₂O₅ doped microstructure. There was almost no difference between the doped and undoped microstructure.

(a)   (b)

Figure 6. SEM micrographs of TiO$_2$ disc samples after sintering at 800 °C for 1hr a) Pure TiO$_2$ and b) Nb$_2$O$_5$ doped TiO$_2$.

Figure 7a and b show conductivity measurements in furnace air environment as a function of temperature for samples with different dopants after sintering at 800 °C for 1hr. Conductivity increased with increasing temperature, though the increase was small at lower temperature. The increase in conductivity is due to higher mobility of impurity ions through the grain boundary and defects at higher temperature, because the defect density and related charge will not change significantly at higher temperature. Based on our preliminary conductivity result, if is clear that defect mobility is a significant factor. Most of our electrical conductivity measurements were repeated at least 2 times, but more experiments are in progress.

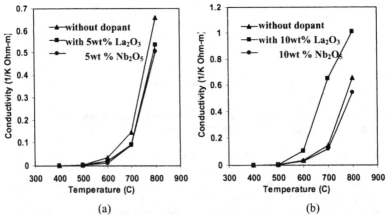

(a)   (b)

Figure 7. Electrical conductivity as a function of temperature for pure and doped TiO$_2$ samples with (a) 5wt and (b) 10wt% dopants.

CONCLUSIONS

Nanostructured TiO_2 based ceramics were synthesized using citrate-nitrate auto combustion method with different dopants. The powder X-ray diffraction data shows that synthesized TiO_2 powders, pure as well as doped, had anatase phase. BET surface area analysis showed a decrease in specific average surface area with increasing dopant concentration. Particle size analysis showed the particle sizes are below 250nm for pure TiO_2 as well as doped TiO_2 samples, and average particle size of doped TiO_2 is decreased. Uniaxially pressed compacts using these nanopowders showed increase in conductivity in the furnace air environment as a function of temperature. It is speculated that the increase in conductivity is due to higher mobility of impurity ions through the grain boundary and defects at higher temperature.

ACKNOWLEDGEMENTS

Authors would like to acknowledge financial support from the NSF under the PECASE grant to Dr. Bose (CTS # 0134476). Authors would also like to acknowledge experimental help from Mr. Ashis Banerjee at WSU. Authors would also like to acknowledge suggestion and helpful discussion with Prof. S. A. Akbar at Ohio State University.

REFERENCES

[1]S. Krathong, C. Saiwan, P. Ouraipryvan, and E. A. O'Rear, "Nano-Titanium Dioxide Synthesis in AOT Microemulsion System with Salinity Scan," *Asian Pacific Confederation of Chemical Engineering*, 9th **#444** (2002).

[2]H.-M. Lin, C.-H. Keng, and C.-Y. Tung, "Gas-Sensing Properties of Nanocrystalline TiO_2," *NanoStruct. Mater.* **9**, 747-750 (1997).

[3]M. Pechini, "Method of Preparing Lead and Alkaline Earth Titanates and Niobates and Coating method Using the same to form a Capacitor," U.S. Pat. No. 3 330 697, July 11, 1967.

[4]Paul A. Lessing, "Mixed-Cation Oxide Powders via Polymeric Precursors," *Ceramic Bulletin,* **68** [5] 1002-1007 (1989).

[5]A.P. Maciel, P.N. Lisboa-Filho, E.R. Leite, C.O. Paiva-Santos, W.H. Schreiner, Y. Maniette and E. Longo "Microstructural and Morphological Analysis of Pure and Ce-doped Tin Dioxide Nanoparticles," *J. European Ceramic Society* **23** [5] 707-713 (2003).

[6]T. Seiyama, A. Kato, K. Fujiishi, M. Nagatini, *Anal. Chem.* **34**, 1502 (1962).

[7]N. Taguchi, Brit. Patent, 1282993, 1288009, 1280803 (1970).

[8]M. Radecka, K. Zakrzewska and M. Reskas, "SnO_2-TiO_2 solid solutions for gas sensors," *Sensors and Actuators B*, **47**, 194-204 (1998).

[9]P.K. Dutta, A. Ginwalla, B. Hogg, B.R. Patton, S. Akbar, etc., "Interaction of Carbon Monoxide with Anatase Surfaces at high Temperatures: optimization of a carbon monoxide sensor," *J. Phys. Chem. B*, **103**, 4412-4422 (1999).

[10]N.O. Savage, S. Akbar, P.K. Dutta, "Titanium Dioxide Based High Temperature Carbon Monoxide Selective Sensor," *Sensors and Actuators B*, **72**, 239-248 (2001).

[11]A.M. Azad, S.A. Akbar, and L.B. Younkman, "High-Temperature Immittance Response in Anatase-Based Sensor Materials," *J. Am. Ceram. Soc.*, **77** [12] 3145~52 (1994).

[12]G.J. Li and S. Kawi. "High-Surface-Area SnO_2: A Novel Semiconductor-Oxide Gas Sensor" *Materials Letters* **34**, 99-102 (1998).

[13]M. Ferroni, M.C. Carotta, V. Guidi, G. Martinelli, F. Ronconi, O. Richard, D.V. Dyck and J. V. Landuyt "Structural Characterization of Nb–TiO_2 Nanosized Thick-Films for Gas Sensing Application", *Sensors and Actuators B: Chemical*, **68**, Issues 1-3, 25, 140-145 (2000).

[14]M. Ferroni, G. Faglia, V. Guidi, P. Nelli, G. Martinelli and G. Sberveglieri, "Characterization of A Nanosized TiO_2 Gas Sensor," *Nanostructured Materials*, **7**, 709-718 (1996).

[15]K.H. Yoon, J. Cho, and D.H. Kang, "Physical and Photoelectrochemical Properties of The TiO_2-ZnO System," *Materials Research Bulletin*, **34** [9] 1451-1461 (1999).

[16]M.C. Carotta, M. Ferroni, D. Gnani, V. Guidi, M. Merli, G. Martinelli, M.C. Casale, M. Notaro, "Nanostructured Pure and Nb-doped TiO_2 as Thick Film Gas Sensors for Environmental Monitoring," *Sensors and Actuators B*, **58**, 310-317 (1999).

[17]S.A. Akbar, L.B. Younkman, and P.K. Dutta, "Selectivity of an Anatase TiO_2-Based Gas Sensor," 160~167 in Polymer in sense, Edited by N. Akmal and A.M. Usmani, Ch. 14, (1998).

[18]S.-F. Wang, Y.-F. Hsu, R.-L. Lee, Y.-S. Lee, "Microstructural Evolution and Phase Development of Nb and Y Doped TiO_2 Films Prepared by RF Magnetron Sputtering," *Applied Surface Science*, **229** [1-4] 140-147 (2004).

[19]M.-H. Wang, R.-J. Cuo, T.-L. Tso, and T.-P. Perng, "Effects of Sintering on the Photoelectrochemical Properties of Nb-Doped TiO_2 Electrodes," *Int. J. Hydrogen Energy*, **20** [7] 555-560 (1995).

[20]K.C. Song, J.H. Kim, "Preparation of Nanosize Tin Oxide Particles from Water-inOil Microemulsions," *J. Colloid and Interface Science*, **212**, 193-196 (1999).

[21]D. A. Bonnel, B. Huey and D. Carroll, "In-situ measurement of electric fields at individual grain boun daries in TiO_2," *Solid State Ionics*, **75**, 35-42 (1995).

Nanostructured Membranes, Films, Coatings, and Self-Assembly

SYNTHESIS OF NANOSTRUCTURED OXIDE FILMS VIA CHEMICAL SOLUTION DEPOSITION, MOLECULAR DESIGN, AND SELF-ASSEMBLY

Michael Z. Hu[*]
Separations and Materials Research Group
Oak Ridge National Laboratory
Bldg. 4500N, MS-6181, Oak Ridge, TN 37931, USA
[*]E-mail: hum1@ornl.gov, Phone: 865-574-8782

ABSTRACT

An overview on chemical solution deposition of oxide films is presented in regarding to nanostructure-engineering approaches and their promise in producing nanocrystalline, nanoporous, and mesoporous films or membranes. Results of our deposited films are provided to illustrate the principle that both *molecular engineering* (via molecular design and self-assembly) and *process engineering* at either low-temperature synthesis or high-temperature processing conditions are critical to control the final nanostructure in films (e.g., nanoscale grain or pore size/orientation). Fundamental issues in two general categories of film synthesis, *in-situ growth* and *precursor coating* methods, have been discussed. Nanostructure control in films can be achieved by tuning chemical solution chemistry, by controlling nucleation and growth, and by taking advantage of molecular self-assembly on substrate surface or during drying of solutions containing molecular templates. It is shown that *in-situ* growth in low-temperature solutions could produce desirable nanocrystalline phases in films. Meanwhile, thermal processing as a necessary step for the coating approach could also significantly affect the final nanostructure of films.

INTRODUCTION

There is a great demand for capability to fabricate a uniform layer of nanostructured film with controlled grain size, density, or pore structure and excellent adherence to a solid substrate surface. Oxide ceramic films, such as zirconia, hafnia, silica, and titania, have found significant applications in corrosion protection, sensors, insulating film for capacitors, inorganic membranes for gas separations and catalysis, and in micro-, nano-, and opto-electronic devices. The methods for fabricating oxide thin films include a variety of techniques, such as reactive e-beam processes, galvanostatic oxidation, dual ion beam sputtering, and atomic layer epitaxy. These methods typically require high temperature and vacuum. Alternatively, soft chemical solution methods are promising for tailored synthesis (via growth or coating) of nanostructured films, which may contain nanocrystal grains of controlled size or nanoporous/mesoporous structures of controlled size, shape, or orientation.

Soft solution processing has been considered as a special strategy for one-step processing of advanced inorganic materials.[1] One particular approach of using organic molecular self-assembly to enable synthesis of inorganic nanostructures in solutions has a biomimetic origin.[2-11] In nature, many inorganic mineral materials, such as mollusk shells, avian eggshells, teeth, and bones, can be formed by biomineralization (i.e., exquisite control over inorganic solid nucleation, crystallization, and growth at organic interfaces) in simple aqueous solutions at ambient conditions with respect to temperature, pressure, and atmosphere. Usually, high degree nm pore diameters) in some inorganic ceramic nanostructured films. However, technical challenges still remain in the control of mesopore orientation and the achievement of large-area of ordering of the pore arrays. Basic understanding of inorganic-organic hybrid molecular self-assembly phenomena with controlled sol-gel coating process could lead to more advanced adjustment and control of nanostructures such as nanoscale pore size, shape, ordering, and even orientation of the pore channels.

One example of molecular design-based templating of nanostructured coatings is shown in Figure 9. By conventional air drying of mixture of metal alkoxide and P123 in ethanol, films containing finger-print or parallel patterns are typically produced. Such pattern formation is due to self-assembling microphase separation phenomena of block copolymer mixture with inorganic precurosors. These parallel mesoporous channels are randomly formed and basically lying down on the substrate surface (Figure 9). However, oriented mesopore channel structures can be formed by process engineering with controlled assembly. Figure 10 clearly indicates the mesopore channels (7.5 nm diameter) oriented perpendicular to the substrate surface. Further more, large area (up to 3 mm dimension) of hexagonal-phase ordering has been demonstrated.[34]

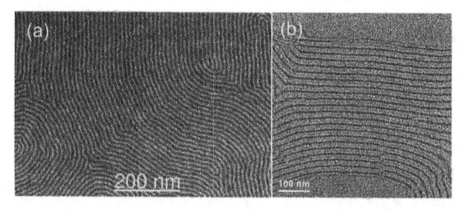

Figure 9. Self-assembled films on substrate surface after natural drying of mixed alkoxide and block copolymer P123 in ethanol. (a) STEM image of a silica film, bright lines are silica rich area, 7.5 nm pore channel diameter . (b) TEM image of a titania film, black lines are titania rich area, ~ 10 nm pore channel diameter.

Thermal processing of zeolite membranes and YSZ nanocrystalline films
The synthesis processes that involve molecular design and engineering and process control directly affect the nanostructure formation in the obtained films. However, post-synthesis processes such as heat treatments are sometimes necessary to crystallize the amorphous precursor films, remove organic precursor elements, or generate desirable pore structures.

We have recently developed a novel method in preparation of top-layered zeolite membranes.[30] The method requires thermal treatment of pre-coated gels on the support, during which the gels are converted into inter-grown zeolite crystal layer on top of a substrate.

advantageous features of the *in-situ* growth approach. It is suitable for uniform layer growth on complicated shapes or large surfaces of solid substrates. The desired surface nucleation or nanocluster deposition affinity to the substrate surface can be induced either by *molecular design*

(a) In-Situ Growth　　　　　　　(b) Coating Approach

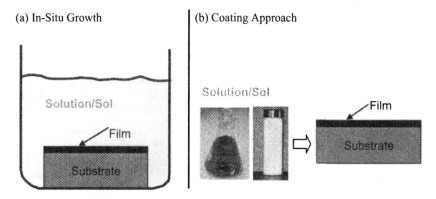

Figure 1. Two types of chemical solution deposition approaches for creating layer of oxide films on a solid substrate surface.

of solution chemistry or by *modification of substrate surface chemistry with self-assembled monolayers* (SAMs). Patterned films with micro- or nanoscale feature resolution can be made via growth/deposition of films on patterned SAMs.[16-20] It makes possible to grow film layers on small curved surfaces of nanoparticles.[21,22] Such coated nanoparticles are important precursor powders for fabricating nanocomposites containing *nanoscale homogeneity* in composition and structure. This approach offers the opportunity of soft chemical solution processing (i.e., low-temperature crystallization during film growth) and thus, avoids the post-synthesis high-temperature calcinations/sintering that are typically required for ceramic processing.

The coating approach usually involves three separate stages: (1) Preparation of a precursor solution or sol, which involves homogeneous nucleation and growth in a solution of designed chemistry. (2) Coating the precursor solution/sol onto the substrate surface via dip- or spin-coating techniques. (3) High-temperature drying and processing (calcinations/sintering). It is important to point out that molecular engineering of inorganic nanostructures can also play a significant role in this coating approach. Besides inorganic chemistry tailoring, organic molecular design elements can be incorporated into the preparation of precursor solution, so that the nanostructure (e.g., pore size, ordering, or orientation) could be controlled.

Four specific experimental film deposition systems are described below:
Hafnia/Zirconia Films by In-situ Growth/Deposition on SAMs. As illustrated in Figure 2(a), SAM layer modifies the substrate surface chemistry and thus induces desirable growth and deposition of films. A few review articles have summarized recent progress in ceramic thin film deposition on SAMs from solutions.[23-25] Organic SAMs, such as those described by Agarwal et al.,[26] are used in our hafnia film deposition studies. They are highly ordered two-dimensional

arrays of long-chain hydrocarbon molecules [X-(CH$_2$)$_n$-Y] of a specific length, which are covalently attached to a substrate through an X-end functional group (such as siloxane bond by reaction of -SiCl$_3$ with OH-covered substrate) and possess a functional surface group Y (such as sulfonate –SO$_3^-$) that is projecting away from the substrate surface. The highly ordered and close-packed characteristics of the monolayer are a result of the strong interactions between the substrate and the monolayer in addition to the short-range van der Waals forces between the chains. The functional terminal group Y on the SAM surface can be chemically modified without disturbing the monolayer to provide a favorable surface functionality necessary to initiate and promote deposition of metal oxide film from the surrounding solutions/colloidal suspensions. In addition to the ability to provide a desired surface functionality, SAMs can withstand temperatures up to 100°C and solutions that are strongly acidic or basic. Readily prepared on a large scale, SAMs are mechanically and solvolytically stable. With these characteristics, SAMs

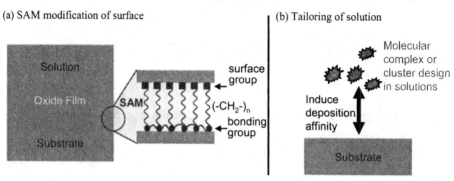

Figure 2. Film deposition on substrate surfaces can be favorably induced by (a) modification of substrate surface chemistry via SAMs or alternatively by (b) tailoring solution chemistry via molecular complex and nanocluster design.

can be used in a variety of conditions for the deposition of various oxides from aqueous or some non-aqueous solutions. The thickness of the SAM layer [-O-Si-(CH$_2$)$_{16}$-SO$_3^-$] is around 25 Å. The distance between the two terminal (thioacetate or sulfonate) groups is 4.3 Å.[26] The chemical conversion of surface terminal group from thioacetate to sulfonate can be done in situ. The deprotonation of the desired sulfonate terminal group (–SO$_3$H → –SO$_3^-$ + H$^+$) could occur even in acid solutions, thus providing a high, uniform negative charge density on the SAM surface. The sulfonate terminal groups (–SO$_3^-$) are believed to be the active sites on which inorganic nanoclusters nucleate or adhere.

Nanocrystalline titania films and zeolite membranes by in-situ growth on substrate surfaces. When the bare silicon wafers are submerged in an acidic solution of titanium tetrachloride adjusted with hydrochloric acid, there was no deposition of film observed. The strategy we use here is to design appropriate titanium complex species in solutions, which induce hydrothermal growth of titania films (Figure 2(b)) on the surface of a submerged substrate (with 10-100 mM TiCl$_4$ and 50-150°C). In the case of zeolite (silicalite-1) membranes, alumina supports were used as substrates. Zeolite top layers are spontaneously grown on substrate surfaces under hydrothermal conditions in NaOH solutions containing dissolved silica and templating molecules such as tetra propyl ammonium hydroxide (TPA).[27]

Nanostructured mesoporous titania and silica films. These films are produced by spin coating of mixtures of metal alkoxide and triblock-copolymers (P123) in alcohols. The process is similar to conventional sol-gel coating, except that molecular self-assembling agents (i.e., P123) are introduced in the inorganic precursor solution and play a templating role to form desirable mesoporous oxide structures on substrate surfaces.

Nanocrystalline yttrium stabilized zirconia (YSZ) films. An amorphous layer is produced by spin coating of polymer precursors, followed by heat treatment in a furnace.[28]

Films are characterized by atomic force microscopy (AFM), scanning electron microscopy (SEM), transmission electron microscopy (TEM), scanning tunneling electron microscope (STEM), X-ray diffraction (XRD), and optical ellipsometry. In the *in-situ* film growth/deposition processes studied here, there is always a solution precipitation coexisting with the film formation process. Therefore, dynamic light scattering (DLS) and small angle X-ray scattering are useful to monitor the homogenous nucleation and growth of particles in the bulk solutions. On the other hand, XRD is a tool to monitor the nucleation and growth of nanocrystals in the precursor films during heat processing.

RESULTS AND DISCUSSION
In the context of this paper, "deposition" means the process of accumulating solid materials of interest on the surface of a substrate from liquid-phase solutions or sols (i.e., colloidally stable suspensions of solid nanoparticles or polymeric nanoclusters). Such deposition can be achieved either by *in-situ growth in solutions* or *solution coating* approaches.

Experimental results of several oxide films by a few different chemical solution deposition methods are briefly summarized here, including (1) hafnia/zirconia film deposition on SAM-modified silicon wafer surfaces, (2) hydrothermal growth of nanocrystalline titania film growth and zeolite films, (3) coating of nanostructured mesoporous titania and silica films, and (4) coating of nanocrystalline yttrium stabilized zirconia (YSZ) films. The first two examples belong to in-situ growth approach and examples of mesoporous films and nanocrystalline YSZ films belong to the coating approach. These examples illustrate that *molecular engineering* (i.e., molecular design, solution chemistry tailoring, and molecular self-assembly) and *process engineering* (i.e., synthesis condition control, drying, calcinations/sintering) play an equally important role in determining the final nanostructure of the deposited film.

Under the conditions of film deposition by *in-situ* growth approach in our work, homogeneous nucleation and precipitation in bulk solutions always co-exist with the film deposition process. Of course, for the coating approach, a precursor sol must be prepared via homogeneous nucleation and growth of nanoclusters or nanoparticles in the solutions. Therefore, understanding of the fundamental nucleation and growth phenomena in liquid solutions becomes relevant to the film deposition by both approaches.

Depsition of hafnia (HfO₂)/zirconia (ZrO₂) films on SAMs

An promising route for the synthesis of oxide thin films involves the use of aqueous chemical solutions and self-assembled molecular monolayers (SAMs) — a biomimetic or bio-inspired process. Due to the chemical similarity between hafnium and zirconium, we believe that the film deposition approaches that have been investigated for zirconia via SAMs may be very well adapted to grow hafnia films. For the first time hafnia films have been successfully grown on the

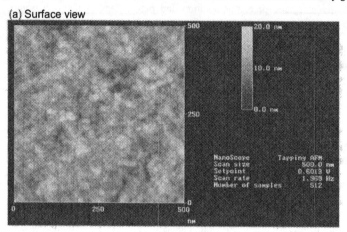

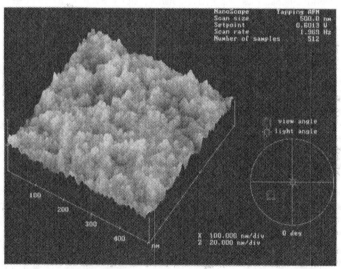

Figure 3. AFM images of a typical hafnia film surface.

SAM-modified silicon wafer surfaces. A typical AFM image of the hafnia film surface is shown in Figure 3. This is indeed similar to the film surface of zirconia as reported previously.[23-26] Nanocluster sizes in the hafnia film are less than 10 nm. The film surface roughness is estimated ± 1 nm. The thickness can be from a few nanometers to hundreds of nanometers depending on the conditions. As compared with the deposition on bare silicon wafer, SAM surface truely induces uniform deposition and improves the film quality in terms of uniformity of thickness and roughness.

We also observed that the precipitated solid particles collected in the bulk synthesis solutions (Figure 4) are very different from the constitutive nanoclusters in the film, indicating that film growth mechanism are either due to heterogeneous *surface ion-by-ion growth* and/or by *nanocluster adherence* to the surface. Only small nanoparticles (<10 nm) appear to have strong affinity to the surface and contribute to the formation of the film, while larger micrometer size colloids as shown in Figure 4 grow and aggregate mainly in the solutions. This observation is in agreement with earlier findings by Supothina et al.[31]: once the diameters of the colloid particles exceed ~10 nm, the forces that would bind them

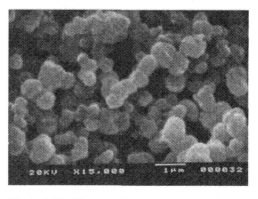

Figure 4. SEM images of a typical precipitated solid particles in bulk solutions collected during hafnia film growth.

to the growing film become negligibly small. The driving force for nanocluster deposition could be described by a coagulation model based on the classical DLVO theory.[32] Nanoparticle size in the film depends on synthesis conditions and is generally larger when there are faster hydrolysis/condensation conditions (i.e., higher salt concentration, higher temperature, lower HCl concentration). The large colloidal particles present in the bulk solution at longer reaction times (as shown in Figure 4) are essentially "inert" (although there is incidental adsorption to the film surface) and thus do not contribute to the growth of the film thickness. Colfen et al.[33] also pointed out the importance of an initial growth stage to zirconia film formation after comparing the initial growth rates of particles and film growth rates.

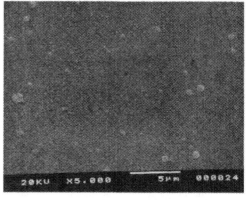

Figure 5. SEM image of a typical surface of titania film by chemical solution growth approach.

This success in depositing hafnia film could open up great opportunities for hafnia-related processing and applications in gas sensors, optical interconnect circuits (such as nanowaveguides), optoelectronic devices, optical coatings (in the visible and IR regions), electronic devices [insulating films (E_g = 5.5 eV) for capacitors or spin-dependent tunnel junctions], catalysis, structural/thermal coatings, corrosion-resistant coatings, and separation membranes.

Titania film growth on unmodified substrate surface

As an alternative approach to the substrate surface chemistry modification such as by SAMs, film deposition from initially clear solutions can be induced by tailoring the solution chemistry or the type of ionic species in solutions. When the bare silicon wafers are submerged in the solutions of plain acidic solution of titanium tetrachloride, no film deposition occurred on the substrate surfaces due to lack of interaction of hydrolyzed titanium species with the substrate. However, when an appropriate titanium complex is formed in the solution, the hydrothermal heating process induced coherent growth of titania films on the substrate surfaces. Various colorful films have been obtained by varying film thickness (green ~ 260-320 nm, gold yellow ~ 188 nm, purple ~ 204 nm) and by varying nanocrystal size inside the films. Figure 5 shows an SEM image of a typical surface of titania films grown by the complexation method. The film is consisted of small nanoparticles (background smooth surface), however, a few adsorbed/trapped large colloids of micrometer size are typically observed on the film surface at the later stage of the film growth. By controlling the precipitation (nucleation and particle growth) in bulk solutions, the undesirable colloids adsorption/attachment can be avoided.

(a)   (b)

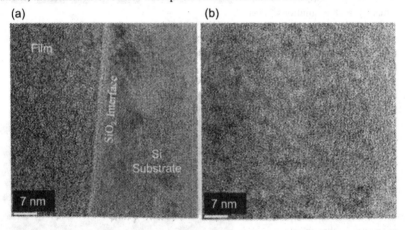

Figure 6. High resolution TEM images of cross section of titania film, showing (a) film-substrate interface and (b) details of nanocrystalline structure inside the film.

Further examination of titania films by TEM and XRD indicates the nanocrystalline structure in the film (Figure 6 and 7). Figure 6 (b) shows the existence of ~3 nm nanocrystals uniformly dispersed inside somewhat structureless film background material. This observation is agreeable with the XRD data showing peak broadening (Figure 7), an indictor of nanocrystalline phases. In contrast to the conventional sol-gel process which usually requires high temperature calcinations to crystallize the coated amorphous precursor films, the as-prepared titania film here prepared at low-temperature (80-100°C) conditions already produced nanocrystalline structure. In addition, XRD data show that lower synthesis condition (at 80°C) produced pure nanocrystalline anatase phase while higher temperature (100°C) condition generated some rutile nanocrystalline phase co-existing with the anatase phase.

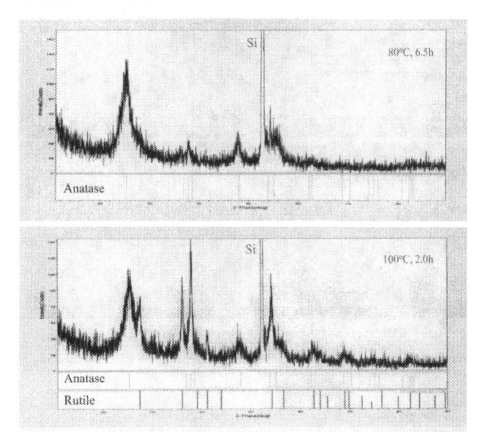

Figure 7. X-ray diffraction (XRD) analysis of titania films (a) obtained at lower synthesis temperature and (b) obtained at higher synthesis temperature.

As illustrated in Figure 8, microcrack formation during drying of as-prepared film can be prevented and features of titania film surfaces can be tailored by tuning chemical solution synthesis conditions. Deposition on vertical surface (Figure 8d) produce finer film surface by avoiding settled particles on the film surface during deposition on horizontal surface (Figure 8b).

Further more, our optical ellipsometry measurements have determined a high refractive index (up to 2.2) and a large band-gap (up to 3.7 eV) for these titania films.

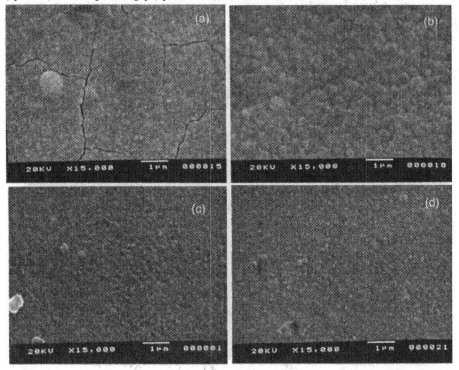

Figure 8. SEM images of titania films with various surface features tailored by synthesis conditions. (a) 90°C, 3h, color green. (b) 80°C, 6h, color green. (c) 90°C, 1h, color blue. (d) 80°C (vertical deposition), 7h, color green.

Engineering of mesoporous nanostructures in oxide films by molecular templating and controlled self-assembly processes

Designed molecular templating approach has become fundamentally important to engineering the porous nanostructures of advanced inorganic materials. A well known example is zeolite (silicalite-1) with highly ordered nanoporous structure created by molecular templating (TPA – tetra propyl ammonium hydroxide).[29] Such nanostructured porous film can be grown by hydrothermal synthesis in caustic solutions containing dissolved silica and TPA.[27] However, nanostructures with pore size larger than 2 nm are very difficult to obtain by this hydrothermal templating synthesis approach.

The sol-gel processing in combination of surfactant or polymeric self-assembly as organic templates offers an opportunity to create large sized pores (i.e., mesopores, ranging from 2 to 50 nm pore diameters) in some inorganic ceramic nanostructured films. However, technical challenges still remain in the control of mesopore orientation and the achievement of large-area of ordering of the pore arrays. Basic understanding of inorganic-organic hybrid molecular self-assembly phenomena with controlled sol-gel coating process could lead to more advanced adjustment and control of nanostructures such as nanoscale pore size, shape, ordering, and even orientation of the pore channels.

One example of molecular design-based templating of nanostructured coatings is shown in Figure 9. By conventional air drying of mixture of metal alkoxide and P123 in ethanol, films containing finger-print or parallel patterns are typically produced. Such pattern formation is due to self-assembling microphase separation phenomena of block copolymer mixture with inorganic precurosors. These parallel mesoporous channels are randomly formed and basically lying down on the substrate surface (Figure 9). However, oriented mesopore channel structures can be formed by process engineering with controlled assembly. Figure 10 clearly indicates the mesopore channels (7.5 nm diameter) oriented perpendicular to the substrate surface. Further more, large area (up to 3 mm dimension) of hexagonal-phase ordering has been demonstrated.[34]

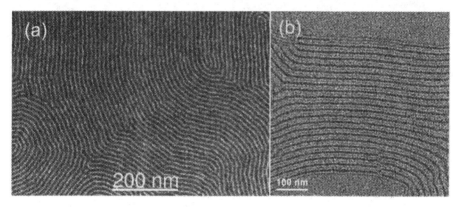

Figure 9. Self-assembled films on substrate surface after natural drying of mixed alkoxide and block copolymer P123 in ethanol. (a) STEM image of a silica film, bright lines are silica rich area, 7.5 nm pore channel diameter . (b) TEM image of a titania film, black lines are titania rich area, ~ 10 nm pore channel diameter.

Thermal processing of zeolite membranes and YSZ nanocrystalline films
The synthesis processes that involve molecular design and engineering and process control directly affect the nanostructure formation in the obtained films. However, post-synthesis processes such as heat treatments are sometimes necessary to crystallize the amorphous precursor films, remove organic precursor elements, or generate desirable pore structures.

We have recently developed a novel method in preparation of top-layered zeolite membranes.[30] The method requires thermal treatment of pre-coated gels on the support, during which the gels are converted into inter-grown zeolite crystal layer on top of a substrate.

In the case of activating zeolite membranes, the organic template molecules trapped in the zeolite cages in the as-prepared films needs to be removed by heat treatment so that nanoporous pores and internal pore surfaces become available. This is commonly known as zeolite pore activation. One important finding from our studies is that programmatic heat treatment and the tailoring solid surface chemistry of the substrate with a suitable transition layer (such as YSZ) is effective to prevent the microcrack formation during the zeolite activation.[27]

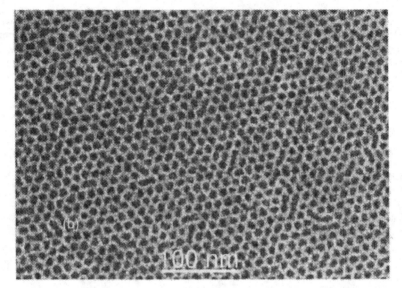

Figure 10. STEM images of oriented mesopore channels in an engineering-assembled silica films on substrate surface. Top surface view of a hexagonally ordered silica film (with black dots corresponding to channels pores of 7.5 nm diameter and bright region to the silica skeleton).

In the case of YSZ film, spin-coating of polymeric precursor solutions first generated an amorphous precursor film. Upon thermal processing, nanocrystallization evolves and further crystal growth occurs with increasing temperature.[30] Here again, knowledge of nucleation and growth phenomena during thermal processing is critical to control the obtained nanocrystalline phase in the YSZ film (Figure 11).

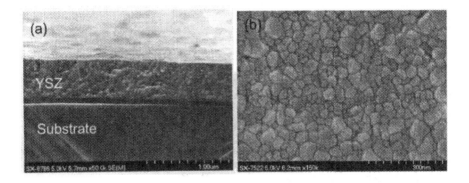

Figure 11. SEM images of YSZ films prepared by the polymer precursor sol-gel coating approach. (a) Cross section view of a film. (b) Surface of a sintered film showing nanocrystal grains.

SUMMARY
Two major approaches for chemical solution deposition of films are discussed, which are *in-situ* growth in solutions and solution coating. Example results presented in this paper illustrate plenty of opportunities in applying molecular design and engineering to tailor solution chemistry and substrate surface chemistry, which are so critical to induce uniform deposition or growth of thin films as well as to achieve controlled self-assembly and templating of desirable inorganic nanostructures. Indeed, soft organic molecules can play an important role in tailoring formation of inorganic nanostructures. In addition, thermal processing plays an equally important role in determining the final nanostructure in inorganic films.

ACKNOWLEDGEMENT
The author would like to thank Y. Wei for the AFM imaging, L. Allard and D. Blom for the TEM and STEM imaging, E. Payzant for the XRD analysis, G. Jellison for optical ellipsometry measurements, J. Dong for the work on nanocrystalline YSZ film, L. Khatri for the zeolite synthesis studies, V. de Almeida for the collaborative mesoporous silica film work, and M. De Guire for the meaningful discussion on SAM-mediated film deposition.

REFERENCES
[1]M. Yoshimura, W. L. Suchanek and K. Byrappa, "Soft solution processing: A strategy for one-step processing of advanced inorganic materials," MRS Bulletin, 25, 17-25 (2000).

[2]J. Bill, R. C. Hoffmann, T. M. Fuchs and F. Aldinger, "Growth of thin ZnO films from aqueous solutions in the presence of PMAA-graft-PEO copolymers," Zeitschrift Fur Metallkunde, 93, 478-489 (2002).

[3]A. A. Campbell, "Interfacial regulation of crystallization in aqueous environments," Curr. Opion Colloid Inter. Sci., 4, 40-45 (1999)

[4]J. Liu, A. Y. Kim, L. Q. Wang, P. J. Palmer, Y. L. Chen, P. Bruinsma, B. C. Bunker, G. J. Exarhos, G. L. Graff and P. C. Rieke, "Self-assembly in the synthesis of ceramic materials and composites," Adv. Colloid Interface Sci., 69, 131-180 (1996).

[5]P. Calvert and P. Rieke, "Biomimetic mineralization in and on polymers," Chem. Mater., 8, 1715-1727 (1996)

[6]I. A. Aksay, M. Trau, S. Manne, I. Honma, N. Yao, L. Zhou, P. Fenter, P. M. Eisenberger and S. M. Gruner, "Biomimetic pathways for assembling inorganic thin films," Science, 273, 892-898 (1996).

[7]P. Calvert and S. J. Mann, "Synthetic and biological composites formed by insitu precipitation," J. Mater. Sci., 23, 3801-3815 (1988).

[8]A. H. Heuer, D. J. Fink, V. J. Laraia, J. L. Arias, P. D. Calvert, D. H. Thompson, A. P. Wheeler, A. Veis and A. I. Caplan, "Innovative materials processing strategies - a biomimetic approach," Science, 255, 1098-1105 (1992).

[9]S. Mann, S., D. D. Archibald, J. M. Didymus, T. Douglas, B. R. Heywood, F. D. Meldrum and N. J. Reeves, "Crystallization at inorganic-organic interfaces - biominerals and biomimetic synthesis," Science, 261, 1286-1292 (1993).

[10]B. C. Bunker, P. C. Rieke, B. J. Tarasevich, A. A. Campbell, G. E. Fryxell, G. L. Graff, L. Song, J. Liu, J. W. Virden and G. L. McVay, "Ceramic thin-film formation on functionalized interfaces through biomimetic processing," Science, 264, 48-55 (1994).

[11]J. H. Fendler, Membrane-Mimetic Approach to Advanced Materials, Advances in Polymers Science Series113, Springer-Verlag: Berlin, 1994.

[12]V. V. Roddatis, D. S. Su, F. C. Jentoft and R. Schlogl, "Temperature- and electron-beam-induced crystallization of zirconia thin films deposited from an aqueous medium: a transmission electron microscopy study," Phil. Mag. A, 82, 2825-2839 (2002a).

[13]V. V. Roddatis, D. S. Su, E. Beckmann, F. C. Jentoft, U. Braun, J. Krohnert and J., R. Schlogl, "The structure of thin zirconia films obtained by self-assembled monolayer mediated deposition: TEM and HREM study," Surf. Coat. Technol., 151-152, 63-66 (2002b).

[14]M. Yoshimura and J. Livage, "Soft processing for advanced inorganic materials," MRS Bull., 25, 12-13 (2000).

[15]S. Supothina and M. R. De Guire, "Characterization of SnO2 thin films grown from aqueous solutions," Thin Solid Films, 371, 1-9 (2000).

[16]Y. Gao, Y. Masuda, T. Yonezawa and K. Koumoto, "Site-selective deposition and micropatterning of SrTiO3 thin film on self-assembled monolayers by the liquid phase deposition method," Chem. Mater. 14, 5006-5014 (2002).

[17]N. Saito, H. Haneda, D. Li, and K. Koumoto, "Characterization of zinc oxide micropatterns deposited on self-assembled monolayer template," J. Ceram. Soc. Jpn., 110, 386-390 (2002).

[18]Y. Masuda, W. S. Seo and K. Koumoto, "Arrangement of nanosized ceramic particles on self-assembled monolayers," Japn. J. Appl. Phys., 39, 4596-4600 (2000).

[19]Z. Zhong, B. Gates, Y. Xia and D. Qin, "Soft lithographic approach to the fabrication of highly ordered 2D arrays of magnetic nanoparticles on the surfaces of silicon substrates," Langmuir, 16, 10369-10375 (2000).

[20]R. J. Collins, H. Shin, M. R. De Guire, A. H. Heuer and C. Sukenik, "Low temperature deposition of patterned TiO2 thin films using photopatterned self-assembled monolayers," Appl. Phys. Lett., 69, 860-862 (1996).

[21]V. Eswaranand and T. Pradeep, "Zirconia covered silver clusters through functionalized monolayers," J. Mater. Chem., 12, 2421-2425 (2002).

[22]Y. Wang, S. Supothina, M. R. De Guire and A. H. Heuer, "Deposition of compact hydrous aluminum sulfate thin films on titania particles coated with organic self-assembled monolayers," Chem. Mater. 10, 2135-2144 (1998).

[23]T. P. Niesen and M. R. De Guire, "Review: deposition of ceramic thin films at low temperatures from aqueous solutions," Solid State Ionics, 151, 61-68 (2002).

[24]T. P. Niesen and M. R. De Guire, "Review: Deposition of ceramic thin films at low temperatures from aqueous solutions," J. Electroceram., 6, 169-207 (2001).

[25]M. R. De Guire, T. P. Niesen, S. Supothina, J. Wolff, J. Bill, C. N. Sukenik, F. Aldinger, A. H. Heuer and M. Ruhle, "Synthesis of oxide and non-oxide inorganic materials at organic surfaces," Z. Mettallkd., 89, 758-766 (1998).

[26]M. Agarwal, M. R. De Guire and A. H. Heuer, "Synthesis of ZrO2 and Y2O3-doped ZrO2 thin films using self-assembled monolayers," J. Am. Ceram. Soc., 80, 2967-2981 (1997).

[27]J. Dong, Y. S. Lin, M. Z. Hu, R. A. Peascoe and E. A. Payzant, "Template Removal Associated Microstructural Development of Ceramic Supported MFI Zeolite Membranes," Microporous and Mesoporous Mater., 34, 241-253 (2000).

[28]J. Dong, M. Z. Hu, E. A. Payzant, T. R. Armstrong and P. F. Becher, "Grain Growth in Nanocrystalline Yttrium-Stabilized-Zirconia Thin Films Synthesized by Spin Coating of Polymeric Precursors," J. Nanosci. Nanotechnol., 2, 161-169 (2002).

[29]L. Khatri, M. Z. Hu and M. T. Harris, "Nucleation and Growth Mechanism of Silicalite-1 Nanocrystal During Molecularly Templated Hydrothermal Synthesis," Ceram. Trans., 137, 3-21 (2003).

[30]J. Dong, E. A. Payzant, M. Z. Hu, D. W. DePaoli and Y.-S. Lin, "Synthesis of MFI-Type Zeolite Membranes on Porous α-Alumina Supports by Wet Gel Crystallization in the Vapor Phase," J. Mater. Sci., 38, 979-985 (2003).

[31]S. Supothina, M. R. De Guire, A. H. Heuer, T. P. Niesen, J. Bill and F. Aldinger, "Deposition of nanocrystalline tin (IV) oxide films on organic self-assembled monolayers," In Organic-Inorganic Hybrid Materials II, MRS Symp. Proc., L. C. Klein, L. F. Francis, M. R. De Guire, J. E. Mark, Eds., Materials Research Society: Warrendale, PA, 576, 203-208 (1999).

[32]H. Shin, M. Agarwal, M. R. De Guire and A. H. Heuer, "Deposition mechanism of oxide thin films on self-assembled organic monolayers," Acta Mater., 46, 801-815 (1998).

[33]H. Colfen, H. Schnablegger, A. Fischer, F. C. Jentoft, G. Weinberg and R. Schlögl, "Particle growth kinetics in zirconium sulfate aqueous solutions followed by dynamic light scattering and analytical ultracentrifugation: Implications for thin film deposition," Langmuir, 18, 3500-3509 (2002).

[34] V. F. de Almeida, D. A. Blom, L. F. Allard, M. Z. Hu, S. Dai, C. Tsouris, Z. Zhang, "Pore Channel Orientation In Self-Assembled Inorganic Mesostructures," poster presentation during Microscopy and Microanalysis 2003, San Antonio, Texas from August 3 through 7, 2003.

GRAIN GROWTH OF NANOCRYSTALLINE Ru-DOPED SnO₂ IN SOL-GEL DERIVED THIN FILMS

Yin Tang and Mark R. DeGuire
Department of Materials Science and Engineering
Case Western Reserve University, Cleveland, OH, 44106

Rocky Mansfield and Nick Smilanich
Sensor Development Corporation, Parma, OH, 44130

ABSTRACT

The microstructural evolution of sol-gel-derived nanocrystalline SnO_2 in thin films, with and without ruthenium doping, was studied. The starting materials were dried gels containing nano-crystals and partially hydrated solids. Crystallization and grain growth were investigated using x-ray diffraction (XRD) at temperatures from 300 °C to 800 °C. Crystallization of SnO_2 occurred rapidly above 700 °C. For equivalent heat treatments, Ru doping suppressed SnO_2 nucleation from the gel materials and slowed down the grain growth up to 700 °C. The doping also reduced the (101) texturing of the thin films.

INTRODUCTION

Tin oxide-based gas sensors are the most widely used commercial gas alarm sensors due to their low weight and cost, small size, simple and robust construction, fast response, and high sensitivity. Devices may be fabricated as dense thin films by evaporation[1,2] or CVD[3,4] or as porous layers by a variety of techniques (sputtering,[5,6] spin-coating,[7,8] or spraying[9,10]).

Sol-gel routes[11-14] allow processing to be carried out at relatively low temperature, and can provide nanocrystalline materials with uniform grain size and homogeneous three-dimensional porosity. These microstructural characteristics have the potential of enhancing the kinetics of surface reactions with detected gases, even at lower operation temperature.

One of the principle disadvantages of tin oxide gas sensors is their lack of long-term stability. It is believed that grain growth with time could be a contributing factor in this regard.[15] The objective of the present work is to study crystallization and grain growth behavior in sol-gel-derived tin oxide with and without Ru dopant.

EXPERIMENTS

Preparation of Pure and Ru-doped SnO_2 Sol-gels

The procedure described here for producing the starting tin oxide sol-gel is based on that reported by Mulvaney et al.[13] Tin (IV) chloride ($SnCl_4$, Aldrich Chemical, 99.995%) (5 ml) was mixed with 250 ml deionized water chilled to 0 °C in an ice-water bath. The solution was stirred using a magnetic stir bar during the whole procedure. White tin oxide gel precipitate was obtained by dropwise addition of 50 ml of 5% aqueous ammonium hydroxide (NH_4OH, Fisher Scientific, reagent grade) solution. Chloride was removed from the precipitate by washing repeatedly in distilled water. A transparent tin oxide sol was obtained by peptizing the precipitate with 5 ml of a 25% ammonia solution for 24 h. The resulting solution had a pH value of about 10. One part by volume of deionized water was added per three parts of solution. Excess ammonium hydroxide was removed by refluxing for about 2 days. The final viscous, transparent, colorless tin oxide sol solution had a pH value around 7. The solution was kept in a refrigerator (4-10 °C) when not in use.

The procedure described here for producing Ru sol-gel is based on that reported by Canevali et al.[16] Ruthenium (III) acetylacetonate (Aldrich) (0.075 g) was dissolved in 50 ml of ethanol-acetylacetone solution (volume ratio 4:1). The solution was stirred using a magnetic stir bar at room temperature for 1 h. Then 10 ml of ethanol-water solution (volume ratio 4:1) were added. Hydrolysis occurs within 2-3 days at room temperature under stirring. The solution can be concentrated to the desired Ru concentration by evaporating more solvents at 60 °C. The resulting transparent red ruthenium oxide solution was kept at room temperature and sealed.

To obtain a Ru/Sn ratio of 1/99, Ru sol was added to the unconcentrated Sn sol (i.e., prior to driving off excess solvent) at 60 °C. After shaking the mixture in a bottle, the color of the solution was uniform. To adjust the viscosity for spin coating, the solution was heated at 60 °C while stirring to evaporate solvent. The doped tin oxide sol-gel was used for spin coating immediately after mixing.

TGA Measurement

On drying at 100 °C for 24 h, the sol-gel solution lost about 98% by weight as water. The resulting white powder was used in thermogravimetric analysis (TGA) (model TGA 2950, TA Inc, New Castle, DE) at a heating rate of 10 °C/min up to 1000 °C.

Thin Film Fabrication

Tin oxide thin films with or without Ru doping were produced by spin-coating (P-6000, Specialty Coating Systems, Indianapolis, IN) sol-gel solution on polycrystalline alumina substrates (13 mm × 15 mm) at 5000 rpm for 40 s. For XRD measurements, to enhance the intensity, 6 layers of sol-gel were applied to each substrate. Each layer was dried at 100 °C for 3 min before the next layer was applied. The final film was dried at 100 °C for 30 min.

The Ru sol was a low-viscosity liquid. One drop was placed on a cleaned {100} single crystal Si wafer (10 mm × 10 mm × 0.5 mm). The sample was heated at 300 °C for 1 h, during which 5 more drops were applied periodically to increase the amount of detected material. The sample was then heated at 800 °C for 30 min.

Grain Growth Measurement

Glancing incidence XRD (GIXRD) measurements were conducted to study the crystal structure of the resulting material. The following parameters were held constant: monochromatic Cu Kα (λ = 0.1542 nm) radiation, tube voltage 45 kV, current 40 mA, incidence angle from 0.5° to 1°, scan rate 2°/min, and scan step 0.02°. This allows changes in the relative amount of crystalline material to be measured from changes in peak intensities (total peak area) between diffraction patterns for the same sample measured at different temperatures.

Hot-stage x-ray diffraction (XRD) (XGEN-4000, Scintag Inc.) was used on thin film samples for in-situ study of the grain growth of tin oxide while heating from room temperature to 800 °C at 2 °C/min. XRD patterns were recorded ($32° \leq 2\theta \leq 40°$) every 10 min during heating. The crystal size was calculated using the Scherrer formula:

$$D = 0.9\lambda / \beta \cos\theta$$

where λ is the wavelength, and β is the corrected half width of the (110) or (101) diffraction peak at half maximum (FWHM), given by

$$\beta^2 = \beta_s^2 - \beta_0^2$$

where β_s is the FWHM of the specimen, and β_0 is the FWHM of an internal standard (here, the Al_2O_3 substrate).

Additional pre-dried thin film samples were heat-treated at 500, 600, 700, and 800 °C for 1 h. These samples were studied using GIXRD at room temperature using the same conditions as described above.

RESULTS AND DISCUSSION
Thermal Decomposition

The TGA trace (Fig. 1) exhibits several distinct weight losses at about 110, 225, and 300-420 °C, and continuous gradual weight loss from 450 to 660 °C. Both the observed weight losses and the transition temperatures exhibited close agreement with the prior study by Giesekke et al.[17] Even though the sol-gel was dried at 100 °C for 24 h, the powder still will adsorb water from the atmosphere. The weight loss at 110 °C is attributed to removal of this physically adsorbed water. The weight loss at 225 °C is assigned to the removal of chemically adsorbed water from SnO_2 particle surfaces. The weight loss between 300 and 420°C is assigned to the removal of OH groups from either the particle surfaces or from oxygen sites.[18] During the last weight loss step, OH groups and H species that are bonded with lattice O or surface O are given off.[19,20]

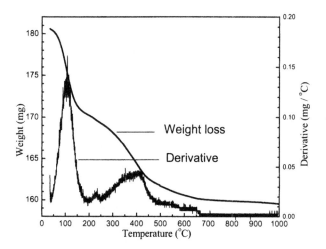

Figure. 1. Thermogravimetric analysis of the SnO_2 sol-gel (after drying at 100 °C).

Tin Oxide Grain Growth

The (110), (101), and (211) peaks of cassiterite (JCPDS #71-0652) were apparent in the XRD pattern from the undoped gel dried at 110 °C (Fig. 2a), although the crystallinity was low. The average crystal size calculated from the peak widths is 3.2±0.2 nm. TEM confirmed the XRD identification and showed the presence of particles 2-3 nm in size (Fig. 3).

Figure 2b presents the evolution of the (101) peak during heating to 800 °C. The peak increased in integrated intensity (indicating that amorphous material was crystallizing) and became sharper (indicating increase in average crystal size). The grain size increased from 3.2 nm to 10 nm (Fig. 4a). Figure 4b shows that the amount of crystalline material increased sharply above 700 °C. This is consistent with the TGA results (Fig. 1), which indicate that the sol-gel was still undergoing transformations to intermediate phases (as indicated by the stepwise weight loss) up to 700 °C. The amount of crystalline SnO_2 would remain low until the decomposition was complete. This would also keep the apparent rate of grain growth low, as new SnO_2 crystals would be nucleating throughout the process

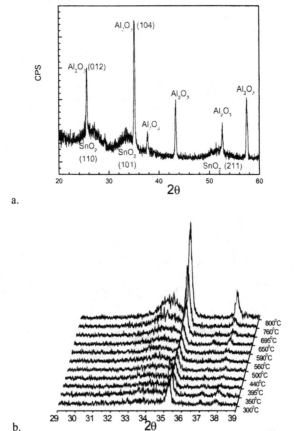

Figure 2. GIXRD spectra of the SnO_2 thin film on Al_2O_3 substrate: a) heat-treated at 110 °C for 0.5 h; b) in-situ measurement while heating in air from room temperature to 800 °C at 2 °C/min.

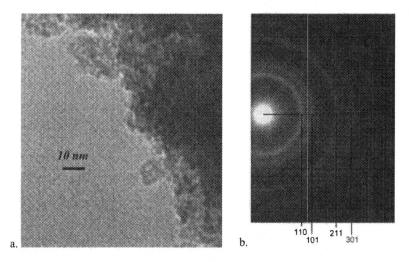

Figure 3. a) TEM image and b) selected area DP of SnO$_2$ powders heat-treated at 110 °C for 1 h.

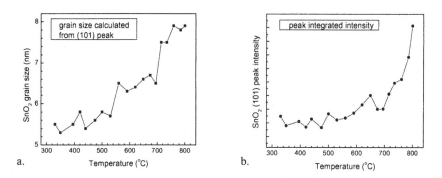

Figure 4. Analysis of XRD data of Figure 2b. a) SnO$_2$ crystal size. b) Total amount of crystalline material, as indicated by the integrated intensity of the (101) peak during heating.

Doping Effects on Tin Oxide Grain Growth

Figure 5 shows the XRD pattern of the Ru sol layers (*i.e.* no SnO$_2$) after heating at 800 °C. All peaks, including the two main peaks at $2\theta=28°$ and 35°, can be attributed to RuO$_2$ (JCPDS #88-0322).

The XRD pattern of the 1 at% Ru-doped SnO$_2$ sol-gel showed only peaks from single-phase cassiterite. The crystal size of the dried Ru-doped SnO$_2$ was 2.4 nm, somewhat smaller than that of the undoped material (Fig. 4a). On heating, the grain growth was noticeably suppressed below 700°C in comparison to the undoped material, and increased from 700 to 800 °C (Fig. 6a). The dependence of peak intensity on temperature (Fig. 6b) approximately paralleled that of the crystal size. This contrasts with the behavior of the pure SnO_2 (Fig. 4b), which showed continuous increase in grain size with increasing temperature, but a rapid increase in peak intensity at about 700 °C. The shallower slope of grain size vs. temperature in Figure 6a compared to Figure 4a suggests that Ru slows the growth rate of SnO_2. While the lack of a sharp increase in total amount of crystalline material above 700 °C in Figure 6b vs. Figure 4b suggest that Ru also suppresses the nucleation rate of SnO_2.

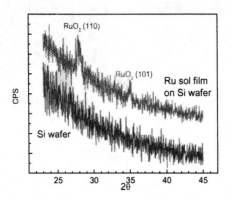

Figure 5. Upper curve: GIXRD spectra of the Ru sol thin film on Si substrate, heat-treated at 800 °C for 0.5 h in air. Lower curve: GIXRD of bare substrate.

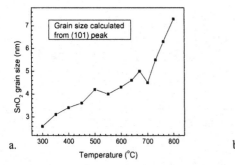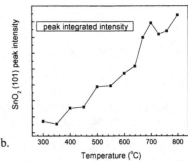

Figure 6. Analysis of XRD data of Ru-doped SnO_2 thin film during sintering in air. a) SnO_2 crystal size. b) Total amount of crystalline material, as indicated by the integrated intensity of the (101) peak during heating.

Ru Doping Effects on Thin Film Surface Orientation

Figure 7a compares the XRD patterns of undoped and Ru-doped SnO_2 thin films heated on Al_2O_3 substrates at 800 °C for 1 h. The main difference between these spectra is the relative intensity of (101) and (110) peaks. In undoped tin oxide powder, per JCPDS file #71-0652, the ratio of the strongest peak (110) to the next strongest peak (101) is 100/75. In the present undoped films, these relative intensities were reversed (Fig. 7b). This suggests that the undoped spin-coated films had partial (101) texturing. In the Ru-doped films, this reversal in intensity was not as strong.

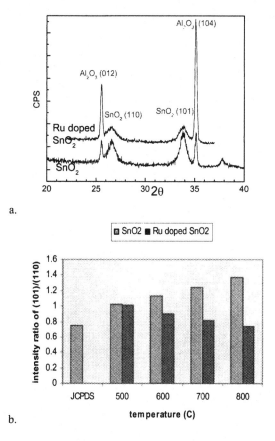

a.

b.

Figure 7. a) GIXRD spectra of the SnO_2 thin films on Al_2O_3 substrate with (bottom curve) and without (top curve) Ru doping, heat-treated at 800 °C for 1 h in air, b) the integrated intensity ratio of (101)/(110) for undoped and doped SnO_2 thin films compared with SnO_2 JCPDS data.

To compare the differences in intensities between the undoped and doped films, the effect of Ru doping on the XRD structure factors must be evaluated. RuO_2 is tetragonal with $a=b=0.4493$ nm and $c=0.3106$ nm. The SnO_2 cassiterite structure is also tetragonal with $a=b=0.4738$ nm and $c=0.3186$ nm. The ionic radius of Sn^{4+} in octahedral coordination is 0.0690 nm while that of Ru^{4+} is 0.0620 nm.[21] The similarities between these two structures and the size of the two ions ($\Delta r = -10\%$) indicates that substitution of Ru for Sn in cassiterite is likely.

The structure factor F_{hkl} represents the amplitude of the (hkl) reflection generated by one unit cell:

$$|F_{hkl}|^2 = \left(\sum_n f_n Cos[2\pi(hx_n + ky_n + lz_n)] \right)^2 + \left(\sum_n f_n Sin[2\pi(hx_n + ky_n + lz_n)] \right)^2$$

where f_n is the atom scattering factor for the atom located at (x_n,y_n,z_n), F_{hkl} is the unit cell structure factor. Table I shows the calculated effects of 1 at% Ru substitution in SnO_2 on the structure factors (F) of the (110) and (101) peaks. (Peak intensity is proportional to the square of the structure factor.) Ru doping decreases the value of $|F|^2$ for both the (110) and (101) peaks, but the effect is negligibly small compared to the intensity differences shown in Figure 7b.

Figure 7b shows the relative peak intensity ratios of (101)/(110) for undoped and doped films after heat-treatment at different temperature for 1 h. The results show that (101) texture for the undoped SnO_2 thin films became enhanced with increasing temperature. The opposite trend is observed for the 1at% Ru-doped materials. With increasing heat treatment temperature, the peak intensity ratios return to the values for SnO_2 powder. The effect of Ru on the structure factors (Table I) is only a small contribution to this effect. Additional contributions may come from effects of Ru doping on the SnO_2 surface energy anisotropy, crystal growth kinetics, nucleation effects or the gel structure itself.

Table I. Atom scattering factors (f) and effects of Ru doping in SnO_2 on XRD structure factor ($|F|^2$).

	(110)	(101)
f_{Sn}	42.6	40.7
f_O	6.2	5.6
f_{Ru}	37.1	35.4
Change in $|F_{hkl}|^2$ after 1 at% Ru doping	Decrease 0.25%	Decrease 0.26%

CONCLUSION

The sol-gel-derived SnO_2 thin films obtained in this work have nanocrystalline cassiterite structure. The starting materials were dried gels containing nano-crystals and partially hydrated solids. Crystallization began during drying at 110 °C, followed with slow continuous crystallization at higher temperatures and a rapid increase after 700 °C in undoped material. Grain size continuously increased with increasing temperature. For equivalent heat treatments, Ru doping suppressed SnO_2 nucleation from the gel materials and slowed down the grain growth up to 700 °C. Ru doping reduced the (101) texture of the thin film with increasing heat treatment temperature, such that at 800 °C the peak intensity relationship returned to the values for SnO_2 powder (i.e. $I_{101}/I_{110} = 0.75$).

REFERENCES

[1]K. D. Schierbaum, J. Geiger, U. Weimar, and W. Gopel, "Specific Palladium and Platinum Doping for SnO_2-Based Thin-Film Sensor Arrays," *Sensors and Actuators B-Chemical,* **13** [1-3] 143-147 (1993).

[2]T. Aste, D. Beruto, R. Botter, C. Ciccarelli, M. Giordani, and P. Pozzolini, "Microstructural Development During the Oxidation Process in SnO_2 Thin-Films for Gas Sensors," *Sensors and Actuators B-Chemical,* **19** [1-3] 637-641 (1994).

[3]L. Bruno, C. Pijolat, and R. Lalauze, "Tin Dioxide Thin-Film Gas Sensor Prepared by Chemical-Vapor- Deposition - Influence of Grain-Size and Thickness on the Electrical-Properties," *Sensors and Actuators B-Chemical,* **18** [1-3] 195-199 (1994).

[4]S. W. Lee, P. P. Tsai, and H. D. Chen, "H-2 sensing behavior of MOCVD-derived SnO_2 thin films," *Sensors and Actuators B-Chemical,* **41** [1-3] 55-61 (1997).

[5]M. C. Horrillo, P. Serrini, J. Santos, and L. Manes, "Influence of the deposition conditions of SnO_2 thin films by reactive sputtering on the sensitivity to urban pollutants," *Sensors and Actuators B-Chemical,* **45** [3] 193-198 (1997).

[6]N. Y. Shishkin, I. M. Zharsky, V. G. Lugin, and V. G. Zarapin, "Air sensitive tin dioxide thin films by magnetron sputtering and thermal oxidation technique," *Sensors and Actuators B-Chemical,* **48** [1-3] 403-408 (1998).

[7]R. S. Niranjan, and I. S. Mulla, "Spin coated tin oxide: a highly sensitive hydrocarbon sensor," *Materials Science and Engineering B-Solid State Materials for Advanced Technology,* **103** [2] 103-107 (2003).

[8]R. E. Cavicchi, R. M. Walton, M. Aquino-Class, J. D. Allen, and B. Panchapakesan, "Spin-on nanoparticle tin oxide for microhotplate gas sensors," *Sensors and Actuators B-Chemical,* **77** [1-2] 145-154 (2001).

[9]M. D. Olvera, and R. Asomoza, "SnO_2 and SnO_2: Pt thin films used as gas sensors," *Sensors and Actuators B-Chemical,* **45** [1] 49-53 (1997).

[10]F. M. Amanullah, K. J. Pratap, and V. H. Babu, "Compositional analysis and depth profile studies on undoped and doped tin oxide films prepared by spray technique," *Materials Science and Engineering B-Solid State Materials for Advanced Technology,* **52** [2-3] 93-98 (1998).

[11]D. J. Yoo, J. Tamaki, S. J. Park, N. Mura, and N. Yamazoe, "Effects of thickness and calcination temperature on tin dioxide sol-derived thin-film sensor," *J. Electrochem. Soc.,* **142**, 1105-1107 (1995).

[12]R. Vogel, P. Hyer, and H. Weller, "Quantum-sized PbS, CdS, Ag_2S, Sb_2S, and Bi_2S_3 particles as sensitizers for various nanoporous wide-bandgap semiconductors," *J. Phys. Chem.,* **98**, 3183-3188 (1994).

[13]P. Mulvaney, F. Grieser, and D. Mesisel, "Electron transfer in aqueous colloid SnO_2 solutions," *Langmuir,* **V6**, 567-571 (1990).

[14]K. Takahata, *"Tin dioxide sensors--development and applications,"* Chemical sensor technology, Ed. Seiyama, T. vol. 1, Elsevier, Tokyo, Japan, p39-55 (1988).

[15]W. Gopel, and K. D. Schierbaum, "SnO_2 Sensors - Current Status and Future-Prospects," *Sensors and Actuators B-Chemical,* **26** [1-3] 1-12 (1995).

[16]C. Canevali, N. Chiodini, F. Morazzoni, and R. Scotti, "Reactivity of SnO_2, Ruthenium-supported SnO_2, Ruthenium (Platinum) ion-exchanged SnO_2 with inert, combustible gases and

air: an electron paramagnetic resonance study," Paper presented at the *Mat. Res. Soc. Symp.*, Boston, US (1999).

[17]E. W. Giesekke, H. S. Gutowsky, P. Kirkov, and H. A. Laitinen, "A proton magnetic resonance and electron diffraction study of the thermal decomposition of tin (IV) hydroxide," *Inorganic Chem.*, **6** [7] 1294-1297 (1967).

[18]D. E. Williams, "Conduction and Gas Response of Semiconductor Gas Sensors," Solid State Gas Sensors, Ed. P.T. Moseley and B.C. Tofield, Adam Hilger, Philadelphia, p86-89 (1987).

[19]W. Dazhi, W. Shulin, C. Jun, Z. Suyvan, and L. Fangging, "Mircostructure of SnO_2," *Phys. Rev. B*, **49**, 14282-14285 (1994).

[20]J. G. Fagan, and V.R.W. Amarakoon, "Reliability and reproducibility of ceramic sensors: Part III, humidity sensors," *American Ceramic Society Bulletin*, **72**, 119-130 (1993).

[21]R. D. Shannon, "Revised Effective Ionic Radii and Systematic Studies of Interatomic Distances in Halides and Chalcogenides," *Acta Cryst.* **A32**, 751-67 (1976).

SELF-ALIGNMENT OF SiO$_2$ COLLOIDAL PARTICLES ON PHYSICALLY AND/OR CHEMICALLY PATTERNED SURFACES

Changdeuck Bae and Hyunjung Shin[*]
School of Advanced Materials Engineering, Kookmin University, Seoul 136-702, Korea

Jooho Moon
Department of Ceramic Engineering, Yonsei University, Seoul 120-749, Korea

ABSTRACT

We report effects of patterned surface structures with the modulation of hydrophobicity in the fabrication of two-dimensional (2D) self-aligned colloidal crystals. Colloidal silica particles have been synthesized by the Stöber process (near monodispersion of 280 nm and 350 nm in diameter). When a drop of the silica suspension with a given volume and concentration (typically 5 μL and 0.5 wt % of the silica particles) is placed upon the substrate, it spreads and forms a concave shape in geometric confinement. In the drop of suspension, horizontal convective flow occurs due to the concave shape and capillary forces. Silica particles are packed together hexagonally resulting in a self-aligned silica monolayer inside the patterned lines. The surfaces with chemical and/or physical patterned lines (~3 and 5 μm in width and spacing) were fabricated on silicon substrate with silicon dioxide by photolithography combined with the modified chemically through SAMs (self-assembled monolayers) of OTS (octadecyltrichlorosilane) using a microcontact printing technique to create alternating hydrophilic/hydrophobic line patterns. Highly ordered silica 2D patterns were observed in the hydrophilic regions, while no silica particles were found in OTS-covered line patterns. Self-aligned silica particle patterns were also observed in a quite large area (at least several hundreds of micrometers) through the line patterns of the surfaces with superhydrophilic modification. This method can serve as a position selective self-alignment aggregation and can be applicable to photonics and micro/nanofabrication.

INTRODUCTION

Colloidal crystals from monodispersed particles can be grown into close-packed three dimensional (3D) and two dimensional (2D)[1] ordered structures with long-range. In recent years one dimensional (1D)[2] crystalline and more complex structures (e.g., body-centered cubic (bcc), simple cubic (sc),[3] dimmer structures,[4] etc.) have been demonstrated as results of continuous efforts. Colloidal crystals having the unique structures are versatile materials not only in the field of photonic crystals but also in nano/microfabrication. When the particles

---

[*] Corresponding author, E-mail: hjshin@kookmin.ac.kr

are assembled together, they have special photonic properties because of the periodic variation of their dielectric constant and thus their potential to show a photonic band gap.[5] Therefore it has been proposed for the next generation pigments and optical devices. For practical applications of photonic crystals with three-dimensional microstructures, it is essential to be able to fabricate patterned crystal films on flat surfaces. It has been known that hydrophobicity plays an important role in the patterning of colloidal crystal films. Many efforts have been devoted to give a contrast in surface wettability. For this purpose, self-assembled monolayers (SAMs) having surface-functionality can provide diverse hydrophobicities. Ultraviolet (UV) irradiation through a photomask is arguably one of the general methods for patterning SAMs.[6-8] Hydrophobic SAMs are coated onto the substrate through an immersion method and then UV irradiation decomposes the exposed SAMs into hydrophilic surface in which colloid crystallization is induced by strong attractive capillary forces. In recent work, the superhydrophilic surface was generated using photocatalytic properties of TiO_2 and it was also observed to exhibit color-change properties.[9]

In this study, we discuss an effect of surface wettability in the patterning of 2D colloidal crystals. For the hydrophobic surfaces, OTS surfaces and hydrogen-terminated surfaces were modified through a microcontact printing technique and chemical etching of Si. Thermally grown as well as chemically oxidized SiO_2 were used to prepare hydrophilic and superhydrophilic surfaces. In addition, we provide results of various modulations of surfaces and identify the most important factor in the application of the patterning of 2D self-aligned colloidal crystals.

EXPERIMENTAL PROCEDURE

Monodispersed silica particles used as self-assembling building blocks were prepared by the Stöber process using tetraethoxyorthosilicate (TEOS, 99.9%, Aldrich), ethanol (99.9%%, Aldrich), NH_4OH (28% NH_3 in water, Aldrich), and deionized (DI) water as starting materials. TEOS was hydrolyzed by a small amount of water in ethanol solutions with ammonia. The reagents were mixed and stirred in a water bath (Fisher isotemp heating circulator model No. 2013P) at 50℃ for 80 min, under N_2 atmosphere. The amount of water and ammonia present in solution regulates the rate of nucleation and growth for the particles and ultimately affects their size.[10] The sol was purified by several cycles of centrifugation, decantation, and resuspension in absolute ethanol to remove undesirable particles and to prevent continuous reaction, followed by drying in a vacuum oven at 100℃. The two different particles populations used in this study had an average diameter of 280 ± 21 nm and 346 ± 53 nm. Measurement of the mean size and size distribution of the prepared SiO_2 colloidal particles were obtained by a light scattering system (Microtrac UPA-150).

Single-side polished p-type Si wafers of [100] orientation were used and 100 nm thick silicon dioxide (SiO_2) as thermally grown. Patterned surfaces were obtained using a photomask with parallel lines (3 and 5 μm in width and spacing, respectively) by photolithography onto the SiO_2/Si substrate.

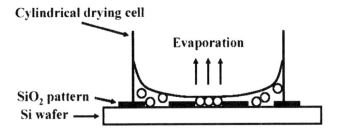

Figure 1. Schematic diagram of the drying cell used to induce the self-aligned 2D colloidal particle assembly onto the patterned substrate.

2D self-assembly of the colloidal SiO_2 suspension on the patterned surfaces was carried out in the drying cell demonstrated by Nagayama and co-workers.[1] A schematic drawing of our experimental set-up for the production of the 2D colloidal crystal is illustrated in Figure 1. The drying cell used a cylindrical plastic tube that geometrically confined the suspension, when glued to the patterned substrate with epoxy. The amount of silica colloids (typically 5 μL with 0.5 wt % suspension) was calculated to cover the whole of the surface. Scanning electron microscopy (SEM, Hitachi S-4300) was employed to image the self-aligned colloidal structures.

RESULTS AND DISCUSSION

To our knowledge, the effect of wettability has not been studied during the growth of colloidal crystals on physically patterned surfaces. For the growth behavior of 2D colloidal crystals on the flat surfaces (i.e. non-patterned substrates), the recent works of Nagayama and co-workers provide comprehensive insight.[11] They also carried out the experiments in a geometrically confined cell using a Teflon ring placed upon the glass plate. A latex suspension was taken into the cell. This particular structure provides a slightly concave shape of the overall liquid surfaces. As the liquid layer thickness gradually decreases due to the water evaporation, they observed that the formation of the 2D colloidal array starts when the thickness of the water layer becomes approximately equal to the particle diameter and crystal growth takes place through a directional motion of particles toward the ordered regions. Considering their experimental observations, they proposed the two-stage mechanism of the ordering process of 2D colloidal crystals. The first stage is dominated by the "immersion capillary force" that when the water level of the thinning aqueous layer in the wetting film fall from the tops of the particles to a low position, namely, when the particles protrude over the water level, the deformation of the liquid-gas interface gives a strong inter-particle capillary forces. However, the transfer of particles from the bulk of the suspension totally immersed toward the ordered array cannot be explained by the action of the capillary force alone. Consequently the mechanism for the second stage suggested "convective transfer" as following: once ordered regions are formed, the capillary pressure due to the curvature of the liquid surface between neighboring particles in that regions increase due to the increase in the curvature of the menisci between them. Therefore, a pressure gradient from the suspension toward the wetting film arises due to the water evaporation and this gradient produces a suspension influx toward the ordered regions.

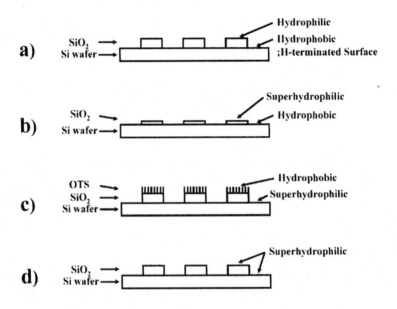

Figure 2. Various surface structures with the modulation of wettability onto the patterned surfaces.

Figure 2 shows different surface structures with the modulation of the wettability onto the patterned surfaces. The first structure was fabricated by conventional photolithography onto SiO_2/Si substrate as shown in Fig. 2(a). SiO_2 thin films without protection with a photoresist (PR) layer were etched in hydrofluoric acid (HF) resulting hydrophobic surfaces. SiO_2 that had been protected through PR has relatively hydrophilic surfaces with a contact angle of 45

± 7° similar to the thermally grown SiO_2. Si surfaces etched with HF provide hydrophobic surfaces having the contact angle of 97 ± 5°. The modulation of superhydrophilic and hydrophobic surfaces as shown in Fig. 2(b) is prepared by the same procedure as Fig. 2(a) and further SiO_2 patterns without a protection layer were exposed in HF for a few seconds. In this case, the surfaces of HF etched SiO_2 present superhydrophilic property showing the contact angle of nearly 0°. To prepare the hydrophobic surfaces we used OTS-SAMs surfaces by a microcontact printing technique.[12] The modulation of hydrophilic and hydrophobic surfaces was prepared as shown in Fig. 2(c). First, the substrate of Fig. 2(a) was chemically oxidized in piranha solution (7:3 concentrated $H_2SO_2:H_2O_2$) at about 100 ℃ for 10 min for the growth of OTS-SAMs. As a result, 2 nm thick chemical oxide films were grown on the overall surfaces of relief structure. Second, microcontact printing was performed using a poly(dimethylsiloxane) (PDMS) stamp only onto the surfaces contacted. The OTS-coated surface presents hydrophobic property of the contact angle of 112 ± 6°, while the regions without OTS were expected to be superhydrophilic surfaces of the contact angle of 0°, due to the hydroxyl surface on chemically grown oxide layer through piranha cleaning. Finally, Fig. 2(d) illustrates modified Fig. 2(a) to superhydrophilic in overall regions through piranha cleaning.

Colloidal suspensions with the final concentration of 0.5 wt % (in 1:1 ethanol:DI water) were dispersed through extensive sonication. A drop spreads in the cell and then assembles together hexagonally from monolayer at the center. During drying, the silica particles together into close-packed assemblies by combination to provide the particles through convective flow from far and lateral capillary forces between near particles, as has previously been described. In the case of Fig. 2(a), the line patterns give rise to liquid flow in the direction parallel to the lines so that well-ordered spheres are found between the lines. Nevertheless, it has been often found that a small number of isolated spheres attaching in the sides of the lines sitting on the top of the patterned SiO_2. To improve the quality of 2D colloidal crystals stripes, not only to make the top surface of relief structure into hydrophobic property through chemical modification, but also various modulations of surface hydrophobicity, and up to the degree without isolated aggregations on the top region of the thin film, several modulation surface structures were designed as shown in Figs. 2(b) – (d). We can observe peculiar result in the patterns of Fig. 2(b) as inversion of Fig. 2(c). The superhydrophilic regions are although higher than the groove regions; the patterns of colloidal crystal were formed upon the surface of SiO_2 thin film (see Fig. 3(b)). We observed relatively clear 2D colloidal stripes without any aggregates of silica colloids in upper surfaces of the thin films as shown in Figs. 3(c) and (d). However, it can be observed remained "branches" of small-size embryo in the sides of the stripe in Fig. 3(c). Figure 3 (d) clearly shows the patterns that cannot observe even any embryo branches.

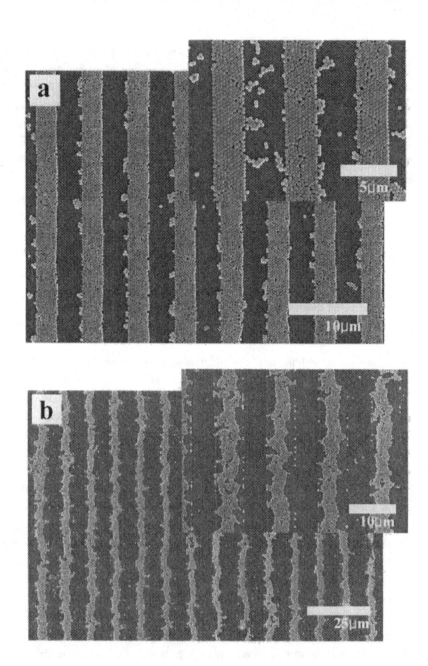

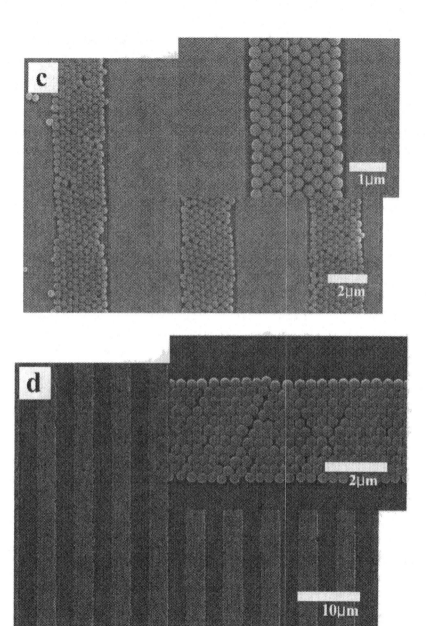

Figure 3. SEM images of 2D self-aligned colloidal particles onto different modulated surfaces corresponding to those in Figure 2(a) – (d), respectively.

For our experimental results, although we cannot yet offer a full explanation because the dynamics of processes were not directly observed, we believe that the wettability effect exceeds the gravitational effect within certain potential energy difference. In our case, the physical relief structure and wettability differences were mixed. First, the relief structure induces the particles to order the groove regions in the patterns because the time that can move increases in the suspension inside walls due to the height difference of silicon dioxide thin film. However, the patterns (Figure 3b) and the aggregates (Figure 3a and 3c) on the top region of the thin film were observed, and hence, they are not subjected to the action of gravitational forces due to the height difference of SiO_2. As a result, the degree of hydrophobic surface is playing an important role over the geometrical confinement.

CONCLUSION

We have demonstrated that the modulation of hydrophobicity can be used to fabricate 2D self-aligned lines of colloidal crystals. Alternating hydrophilic/hydrophobic line patterns through OTS and chemical etching were created and a geometrically confined cell was employed. Therefore, it was achieved 2D strips of colloidal crystals of good quality. These results provide an importance advance towards the application of colloidal crystals.

ACKNOWLEDGEMENT

This work was supported by a grant No. R01-2002-000-00318-0 from Korea Science & Engineering Foundation.

REFERENCES

[1]N. D. Denkov, O. D. Velev, P. A. Kralchevsky, I. B. Ivanov, H. Yoshimura, and K. Nagayama, "Mechanism of Formation of Two-Dimensional Crystals form Latex Particles on Substrates," *Langmuir*, **8**, 3183-3190 (1992).

[2]G. Su, Q. Guo, and R. E. Palmer, "Colloidal Lines and Strings," *Langmuir*, **19**, 9669-9671 (2003).

[3]J. P. Hoogenboom, C. Retif, E. de Bres, M. van de Boer, A. K. van Langen-Suurling, J. Romijn, and A. van Blaaderen, "Template-Induced Growth of Closed-Packed and Non-Close-Packed Colloidal Crystals during Solvent Evaporation," *Nano Letter*, **4** [2] 205-208 (2004).

[4]Y. Lu, H. Xiong, X. Jiang, Y. Xia, M. Prentiss, and G. M. Whitesides, "Asymmetric Dimers Can Be Formed by Dewetting Half-Shells of Gold Deposited on the Surfaces of Spherical Oxide Colloids," H*Journal of the American Chemical Society*H, **125** [42] 12724-12725 (2003)

[5]Cefe López, "Materials Aspects of Photonic Crystals," *Advanced Materials*, **15** [20] 1679-1704 (2003).

[6]Y. Masuda, W. S. Seo, and K. Koumoto, "Arrangement of Nanosized Ceramic Particles on Self-assembled Monolayers," *Japanese Journal of Applied Physics*, **39**, 4596-4600 (2000).

[7]Y. Masuda, K. Tomimoto, and K. Koumoto "Two-Dimensional Self-Assembly of Spherical Particles Using a Liquid Mold and Its Drying Process," *Langmuir*, **19**, 5179-5183 (2003).

[8]C. A. Fustin, G. Glasser, H. W. Spiess, and U. Jonas, "Site-Selective Growth of Colloidal Crystals with Photonic Properties on Chemacally Patterned Surfaces" *Advanced Materials,* **15** [12] 1025-1028 (2003).

[9]Z. Z. Gu, A. Fujishima, and O. Sato, "Patterning of a Colloidal Crystal Film on a Modified Hydrophilic and Hydrophobic Surface," *Angewandte Chemie International Edition,* **41** [12] 2067-2070 (2002).

[10]A. Rugge, J. S. Becker, R. G. Gordon, and S. H. Tolbert, "Tungsten Nitride Inverse Opals by Atomic Layer Deposition," *Nano Latter,* **3** [9] 1293-1297 (2003).

[11]A. S. Dimitrov and K. Nagayama, "Continuous Convective Assembling of Fine Particles into Two-Dimensional Array on Solid Surfaces" *Langmuir,* **12,** 1303-1311 (1996).

[12]Y. N. Xia and G. M. Whitesides, "Soft Lithography," *Angewandte Chemie International Edition,* **37,** 551-575 (1998).

TITANIUM DIOXIDE LOADED ANODIZED ALUMINA NANO-TEMPLATE

Ahalapitiya H. Jayatissa
Mechanical, Industrial & Manufacturing
Engineering (MIME) Department
The College of Engineering
The University of Toledo, OH 43606
E-mail: jayatis66@yahoo.com

Tarun Gupta
Industrial & Manufacturing
Engineering (IME) Department
College of Engineering
Western Michigan University, MI 49008
E-mail:tarun.gupta@wmich.edu

ABSTRACT

Self-assembled titanium dioxide nanoparticles can be used in different applications including catalysts, sensors and biomedical applications. However, the fabrication of such organized titanium dioxide particle array is a challenging problem. In this research, an attempt was made to load the titanium nanoparticles into an anodized alumina template. Aluminum templates with pore size in the range of 35-45 nm were fabricated by the anodization of aluminum plates in an oxalic acid solution. Titanium dioxide nanoparticles were synthesized using a colloidal titanium oxide complex. These nanoparticles were loaded into the alumina template by electrophoresis deposition in an aqueous medium. The loaded templates were investigated using the X-ray photoelectron spectroscopy, scanning electron microscopy and atomic force microscopy measurements.

INTRODUCTION

Nanofabrication by self-organization methods has attracted much attention because of their mass production capability and their capability to overcome limitations in lithography based fabrication methods.[1] The self-organized structures will provide many answers for nano-scale manufacturing which hindered by manipulation difficulties. One approach for assembling of nanostructures is the use of a nanotemplate for organization of high-density material structures. Recently, porous alumina attracted much attention because the size of nanoporosity can be controlled in anodized alumina template.[2] Porous alumina fabricated by electrochemical anodic oxidation of aluminum provides highly ordered arrays of nanoholes in the rage of several hundreds down to the several tens of nanometers in size.[1,2] In this research, we have investigated the deposition capability of dielectric materials in porous alumina template using TiO_2 as the model material.

For this study, TiO_2 nanoparticles were selected because of their many applications in catalytic, gas sensors, biomedical, and photonic devices[3]. Preparation of high-density nanostructured TiO_2 clusters can be useful in those applications in which surface area plays an important role. We have developed the fabrication of template material, preparation of TiO_2 colloids and electrophoresis coating of nanoparticles in the template. The most interesting finding was that the TiO_2 nanoparticles coated in the porous structure selectively due to the effect of electrical field distribution in the nanoporous structure.

EXPERIMENTAL

Nanoporous alumina template was fabricated by electrochemical anodization using an oxalic acid solution as described elsewhere.[1,2] Larger porosity was made by replacing oxalic acid with

H_2SO_4 solution. The TiO_2 nanocrystals were prepared as follows: (i) dissolved pure (99.99%) Ti plate in a 2%NH_4F and 1% HNO_3 acid mixture in room temperature; (ii) the excess amount of NH_4OH solution was added to the Ti solution till colloidal form of TiO_2 were appeared (slight purple color); and (iii) washed the colloids with cold (5 °C) de-ionized water till acid and F⁻ ions were eliminated from the solution. In this process, step (ii) was very critical to achieve nanocrystalline TiO_2 and the temperature and stirring speed as well as the rate of addition of NH_4OH must be controlled.

The solution for electrophoresis coating was prepared using a stabilized TiO_2 colloid in a 0.3M acetic acid solution dispersed with an ultrasonicator. The electrophoresis coating was carried out using the template as the anode and a carbon electrode as the cathode in TiO_2 colloidal solution. The applied voltage was 60 V/cm^2 and the current passed was 500 µA. The deposition was always conducted using a newly prepared TiO_2 colloidal solution to avoid formation of large TiO_2 clusters in the solution. In addition to the above anodic deposition, the template was also used as the cathode to test whether there is any deposition. Another test was carried out by immersing the template in the colloidal solution for a desired period. All samples were thoroughly rinsed with de-ionized water to remove any particle on the surface.

The coated samples were tested with scanning electron microscopy (SEM) and atomic force microscopy (AFM). However, no any difference between coated and uncoated samples was observed. Thus, we have used X-ray photoelectron spectroscopy (XPS) to evaluate our samples under the following conditions. The X-ray source has the following properties: voltage = 15kV; Power = 300w; Source = Mg. The ion gun has the following properties: Beam Voltage = 3kV, Emission Current= 25mA; Ion gun pressure = 15mPa; and sputter area: 10mm.

RESULTS AND DISCUSSION
Figure 1 compares the XPS spectra of anodically and cathodically TiO_2 coated templates along with template, which dip immersed in the colloidal solution.

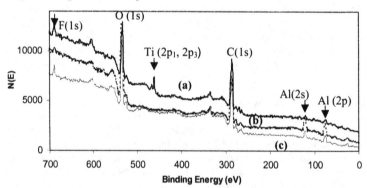

Fig.1. XPS spectra of alumina template (a) electrophoresis coating of TiO_2 using template as the anode, (b) electrophoresis coating of TiO_2 using template as the cathode, and (c) dip immersing of template in TiO_2 colloids.

The XPS spectra indicated that the anodically deposited template has some TiO_2 bands. In an XPS spectrum, Ti $^2p_1$ and Ti $^2p_3$ are exhibited at 458.5 and 464 eV, respectively.[4] In this case, we have seen these bands at 458.3 eV for anodically coated samples. Other two samples did not indicate any signal for TiO_2. This phenomenon indicated that the electrophoresis of TiO_2 colloids results in some deposition in the nanotemplate.

We have also studied the surfaces with SEM and AFM to check whether there is any surface topography variation due to the coating. SEM image a typical template is shown in Fig.2. No any significant change was observed in SEM images for TiO_2 loaded and unloaded templates.

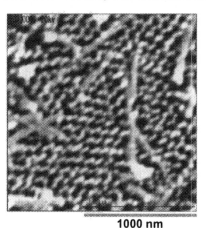

1000 nm

Fig.2. SEM image of nanotemplate fabricated in our laboratories. Some of the lines are due to the unetched Al regions on the plate. No any significant change of surfaces was found before and after deposition of TiO_2 particles.

Due to the above reasons, we have sputtered the samples in XPS measurements for longer periods. Fig.3 indicated the XPS spectra with gradual increase of sputtering time. It can be seen that the template has a clear TiO_2 band with slight increase of intensity with the sputtering time. This was a typical phenomena in XPS measurement in which the peak intensities were increased with sputtering time. Thus, we concluded that the anodic electrophoresis coating resulted in the deposition of TiO_2 in nanoporous regions. As observed in the AFM measurements, the surface topography was not changed due to the coating. These facts suggest that the coating of TiO_2 particles is taken place only in the nanoporous cavities as shown in Fig.4.

The porous alumina templates were fabricated by electrochemical etching of aluminum due to the thickness difference in the surface Al_2O_3 layer of aluminum substrate. Thus, the final finishing of template remains Al_2O_3 walls and holes. However, in this case we have anodized the aluminum plate only partially. Thus, the electrical contacts can be uniformly provided by aluminum plate during electrophoresis coating. When the electric field was applied in the colloidal solution, the field affect only at the cavity regions because of ultrathin Al_2O_3 layers at the bottom of the pores. Other area has much thick (estimated to be 600-800 nm) of Al_2O_3 layers. Thus, electrophoresis coating was taken place only in the porous regions.

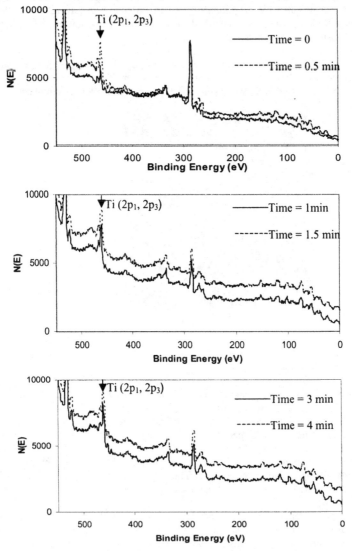

Fig. 3. X-ray photoelectron spectra of electrophoresis coated TiO_2 on alumina template for different sputtering time.

Fig.3. Schematic representation of electrophoresis coated TiO_2 in an alumina template.

In summary, we have demonstrated the coating capability of TiO_2 (or any other dielectrics) materials in the nanoporous regions in anodic alumina. This is a first step investigation for nanofabrication of ceramic structures using a template. We are continuously investigating on this topic to understand related materials issues and application of these types of nanostructures in devices such as gas sensors, biotechnology and catalysts.

ACKNOWKEDGEMNT
One of the authors, AHJ, acknowledges the research funds provided by the University of Toledo, URAF program.

REFERENCES
[1]S. Shoso, "Fabrication of Nanomaterials Using Porous Alumina Templates," *J. Nanoparticle Research*, **5**, 17-30 (2003).
[2]J. Li, C. Papadopoulos, J. M. Xu, and M. Moskovits, "Highly-ordered carbon nanotube arrays for electronics applications," *Appl. Phys. Lett.*, **75**, 367-369 (1999).
[3]S. J. Kim and D. E. Mckean, "Aqueous TiO_2 suspension preparation and novel application of ink-jet printing technique for ceramics patterning," *J. Material Science Letters*, **17**, 141-144 (1997/1998).
[4]D. Ulrike; T. E. Madey, "TiO_2 by XPS," *Surface Science Spectra*, **4**, 227-231 (1996).

FINE-GRAIN NANOCRYSTALLINE TUNGSTEN OXIDE FILMS FOR GAS SENSOR APPLICATIONS

Ahalapitiya H. Jayatissa
Mechanical, Industrial & Manufacturing
Engineering (MIME) Department
The College of Engineering
The University of Toledo, OH 43606
E-mail: jayatis66@yahoo.com

Tarun Gupta
Industrial & Manufacturing
Engineering (IME) Department
College of Engineering
Western Michigan University, MI 49008
E-mail:tarun.gupta@wmich.edu

ABSTRACT

This paper describes the fabrication of fine-grained nanocrystalline tungsten oxide thin films for gas sensor application. Thin films of WO_3 were prepared by thermal evaporation of metallic tungsten oxide powder followed by thermal annealing. The films were annealed at different temperatures to form semiconducting WO_3 films suitable for gas sensor and other electronic device applications. It was found that 300 °C annealed film has a uniformly formed nanocrystalline structure compared with high temperature annealed samples. When these nanocrystalline films were re-annealed at high temperature, the crystallinity and the surface morphology were remained unchanged. In this paper, the structural and surface characteristics of WO_3 films are described as a function of annealing treatments.

INTRODUCTION

Tungsten oxide (WO_3) thin films have been investigated for different electronic device applications including gas sensors.[1] These films were made by high temperature (~600 °C) annealing of vacuum evaporated, sputtered or sol-gel coated WO_3 films. During the high temperature annealing process, the films acquire micro-porous structures hence enhanced gas adsorption properties. Thus, different temperature annealed films exhibited different performances in gas sensors.

Application of annealed WO_3 films in gas sensors has two major problems.[2] (i) High temperature annealed films exhibited larger cracks and high surface roughness. Due to larger cracks, the carrier mobility is drastically reduces hence response time of gas sensors has been increased to a few minutes. Furthermore, patterning of these films in microlithography became difficult because of the high surface roughness of films. (ii) The other major issue of these films is the instability of microstructure at high operation temperatures for long periods. The latter phenomenon is mainly due to the diffusion and coalescence of grains during heating for a long period.

In this study, we have followed two-step annealing of thermally evaporated tungsten oxide to form a stable nanocrystalline WO_3 phase. The first annealing was carried out at a temperature around 300 °C for several hours. Then, we have re-annealed these films at high temperature near 600 °C. The films were characterized using atomic force microscopy (AFM), Raman and X-ray diffraction (XRD) measurements. It was found that the crystallinity and the surface morphology of the nanocrystalline films produced at low temperature were unchanged during high temperature heating. The stability of two-step annealed films indicated that the film preparation method could be a promising way to produce stable WO_3 films.

EXPERIMENTAL

The WO_2 powder was thermally evaporated using a tungsten filament in a high vacuum bell jar system. The substrates of nonalkaline glass and silicon wafers (100 mm) were kept 12 cm away from the filament. The thickness of coated tungsten oxide films was about 350 nm. The films were annealed at different temperatures to identify minimum crystallization temperatures. It was found from X-ray diffraction measurements that the annealing at 300 °C results in the formation of fine-grained nanocrystalline films (as describes in the results and discussion). The coated WO_3 films were annealed in air under following conditions. (i) Samples were annealed at 300 °C for six hours and they were allowed to cool down to the room temperature. (ii) 300 °C annealed samples were re-annealed at 600 °C for 12 hours. In this case, we have also annealed as-deposited samples at 600 °C for comparison. After 12 hour annealing, the samples were allowed to cool down to the room temperature. As-deposited, 300 °C annealed, 300 and 600 °C annealed (two-step), and 600 °C annealed samples were characterized for surface and structural properties in Raman, X-ray diffraction (XRD) and AFM measurements.

RESULTS AND DISCUSSION

Fig. 1 shows the XRD spectra for WO_3 films annealed at different temperatures. It can be seen that the as-deposited films have an amorphous-like structure.[3] We have noticed that these films have high metallic conductivity and light blue color. When these films were annealed over 300 °C for more than five hours, the films changed to the yellowish or transparent films. The conductivity is increased by 3 orders of magnitude during the annealing process. Fig.1 (a) gives the XRD spectra of low temperature annealed films. Even for 200 °C annealing, the crystallinity has been slightly changed. Over 400 °C, polycrystalline formation was observed. As can be seen in Fig.1 (b), 500 and 600 °C annealed films were fully re-crystallized.

Fig. 2 shows the topographs of these films obtained from both AFM and optical microscopy. It can be seen that the as-deposited film has a low surface roughness. When these films were annealed at 300 °C for 5 hours, 50-75 nm size grains were appeared. The optical images indicated that the films were just about to crackdown. When the films were annealed at 600 °C, the lager cracks and some surface roughness were developed in the film.

Judging from the above two observations, we believe that annealing at 300 °C resulted in the formation of a nanocrystalline WO_3 phase. Thus, we have selected 300 °C annealed films for the two-step annealing treatment.

Fig. 3 indicated the Raman spectra of as-deposited, 300 °C annealed, 600 °C annealed and two-step (300 & 400 °C) annealed films. The as-deposited film does not have any characteristics corresponding to a crystalline phase of tungsten oxide.[4] All annealed samples indicated the Raman bands corresponding to monoclinic WO_3. Monoclinic phase is the stable form of WO_3 in room temperature range. 300 °C annealed film and 300/600 °C (two-step) annealed films have very close intensity profiles at the Raman bands of WO_3. The peak width has been reduced unnoticeable amount even though we annealed this film at 600 °C for 12 hours in air. The film directly annealed at 600 °C has the lowest FWHM of the Raman bands, however the difference between 300 °C annealed films and 600 °C annealed films is very small.

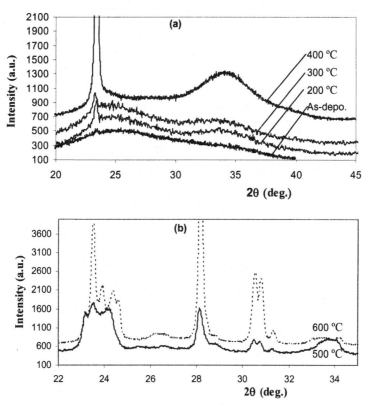

Fig.1. X-ray diffraction of WO₃ films annealed at different temperatures for 6 hours (Annealing temperature is indicated at each curve).

Fig. 4 shows the AFM image of 300/600 °C (two-step) annealed film. It can be seen that even after annealing at 600 °C for 12 hours, the surface roughness was not increased considerably and the larger cracks were not grown. Furthermore, these films have better surface properties and optical transparencies than those made by direct heating at high temperatures. The film adhesion is another important parameter for micromachined gas sensors, which require 2-3 layers of different materials. Thus, the above two-step annealed tungsten oxide films may suitable for wide uses in gas sensor and other optoelectronic device applications.

The advantages of using WO_2 powder are low evaporation temperature and high conductivity of as-deposited films. The high conductive films can be used for electroforming before conversion to semiconducting WO_3 by annealing. As-deposited blue-tungsten oxide films have a resistivity in 10^3 (ohm·cm) range. When these films were annealed around 300 °C, the microstructure of films changed to the fine-grained nanocrystalline oxides. Further increase of temperature resulted in the development of polycrystalline WO_3 films. However, no formation of polycrystalline film was observed by high temperature annealing of nanocrystalline WO_3 (produced by annealing at 300 °C). This result can be interpreted as follows: The as-deposited films, which have very small oxide clusters converted to stable crystalline phases by annealing. The nanocrystalline films do not exhibit recrystallization by re-annealing at higher temperatures because of high stability of nanocrystalline phase. Since nanocrystalline films have high surface area, these films are expected to have better gas sensitivity than polycrystalline films.

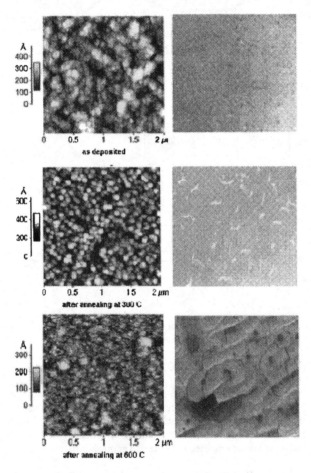

Fig.2. AFM and optical microscope images of films. The treatment conditions are indicated at each micrograph. Magnification of optical images is 200.

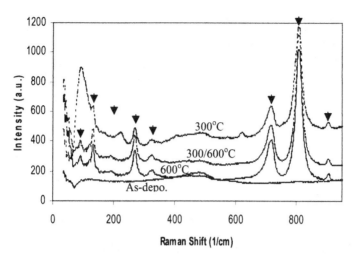

Fig.3. Raman spectra of as-deposited, 300 °C annealed, 300/600 °C (two-step) annealed and 600 °C annealed tungsten oxide films (Substrate: Nonalkaline glass, film thickness: 350 nm). WO_3 peak positions are indicated with arrowheads.

Fig.4. AFM image of 300/600 °C two-step annealed film.

CONCLUSIONS

Systematic investigation was performed to optimize the formation of fine grained nanocrystalline WO_3 films suitable for microfabrication of gas sensors. It was found that the two-step annealing produced stable uniform and low surface roughness films compared with direct annealing at high temperatures. Furthermore, our results indicated that temperature of 300 °C was the critical temperature to form nanocrystalline formation in WO_3 films. All our films were fabricated using commercially available WO_2 powders, which could evaporate at slightly lower temperatures than WO_3. Thus, this material can be used as a low temperature processing material. Further research is underway in our laboratory to improve and use these ceramic thin films in sensor devices.

ACKNOWLEDGEMENTS

One of the authors, AHJ, acknowledges support from the National Science Foundation (Grant #: ECS-0401690) for this research. Research was also supported by the College of Engineering and MIME department at the University of Toledo. They would also like to thank Dr. Toru Aoki at RIE, Shizuoka University, Hamamatsu, Japan for AFM measurements.

REFERENCES

[1]J. J. Ho, "Novel nitrogen monoxides (NO) gas sensors integrated with tungsten trioxide (WO_3)/pin structure for room temperature operation," *Solid-State Electronics*, **47**, 827-830 (2003).

[2]M. Fleischer and H. Meixner, "Fast gas sensors based on metal oxides which are stable at high temperatures," *Sensors and Actuators B: Chemical*, **43**, 1-10 (1997).

[3]B. Ziemer and H. J. Lunkquadif, "A Simple Program for Quantitative X-Ray Powder Diffraction Analysis of Tungsten Oxide and Other Systems with Low Crystallinity," *International J. Refractory Metals and Hard Materials*, **14**, 279-287 (1996).

[4]M. Boulova, N. Rosman, P. Bouvier and G. Lucazeau, "High-pressure Raman study of microcrystalline WO_3 tungsten oxide," *J. of Phys.: Condensed Matter*, **14**, 5849-5863 (2002).

Processing and Characterization
of Nanomaterials

LOW TEMPERATURE CONSOLIDATION OF CERAMIC NANOPARTICLES VIA AN INTERFACIAL ADHESIVE BONDING BY PLASMA POLYMERIZATION

Donglu Shi, Peng He, S. X. Wang*, Wim J. van Ooij, L. M. Wang*
Department of Materials Science and Engineering
University of Cincinnati
Cincinnati, OH 45221-0012

*Dept. of Nuclear Engineering and Radiological Science
University of Michigan
Ann Arbor, MI 48109

ABSTRACT
Using a unique plasma polymerization, the pyrrole-coated alumina nanoparticles were consolidated at a temperature range (~250 °C) much lower than the conventional sintering temperature. The density of consolidated bulk alumina has reached about 95 % of the theoretical density of alumina with an extremely thin polymer coating on the nanoparticle surfaces. After low-temperature consolidation, the micro-hardness test was performed on the bulk samples to study the strength that was related to particle-particle adhesion. High Resolution Transmission Electron Microscopy (HRTEM) experiments showed that an extremely thin film of the pyrrole layer (2 nm) was uniformly deposited on the surfaces of the nanoparticles. The underlying adhesion mechanism for bonding of the nanoparticles is discussed.

INTRODUCTION
The concept of ceramic sintering is based upon the joining of solid particles at interfaces through diffusion at temperatures well above 1000 °C. If these particles can be bonded strongly through other means such as an adhesive coating on the particle surfaces, it may serve as an alternative process to sintering. The low-temperature consolidation will reduce severe surface reactions, geometry distortion, and lower the processing cost since ceramic fabrication requires expensive furnaces and other atmosphere control facilities.[1-6] Furthermore, in the recent development of electronic materials, ceramics are often composed with other types of materials including metals and polymers for device design and achieving unique physical properties. Such a composite cannot be easily processed at high temperatures together due to obvious reasons. It will only be possible to process these composites using a low-temperature consolidation method.

As noted above, one of the possibilities in low-temperature consolidation is through an adhesive layer of polymer thin film between the ceramic particles. In this way, the polymer film acts as a nano-scaled adhesive that bonds the ceramics particles together. This polymer film has to be extremely thin (a few nanometers thick), that makes it fundamentally different from the conventional synthesis where the binder material has a large volume fraction in the ceramic matrix. It will be advantageous to have more contacting surfaces by using nanoparticles so that the consolidation is sufficiently strengthened by more adhesive interfaces in the ceramic matrix.

In this experiment we have attempted to coat a thin adhesive film of polymer onto alumina nanoparticles. After coating the experiments were focused on the consolidation of the coated nanoparticles at a temperature range much lower than the normal sintering temperature. High Resolution Transmission Electron Microscopy (HRTEM) was used to study the extremely thin film of the pyrrole layer (2 nm) on the surfaces of the alumina nanoparticles. Time-of-Flight Secondary Ion Mass Spectroscopy (TOFSIMS) experiments were carried out to analyze the composition of the thin film on the nanoparticles. After low-temperature consolidation, a micro-hardness test was performed on the bulk samples to study the strength that was related to particle-particle adhesion. The underlying adhesion mechanism for bonding of the particles is discussed.

EXPERIMENTAL DETAILS
In this experiment, we selected nanoscale alumina particles ranging from a few nanometers to 150 nm. This large distribution of particles was particularly useful for the study of experimental deposition conditions for different sizes. The schematic diagram of the plasma reactor for thin film deposition of nanoparticles is shown in Fig. 1. The vacuum chamber of plasma reactor consisted of a Pyrex glass column about 80 cm in height and 6 cm in internal diameter.[7,8] The nanoparticles of alumina were vigorously stirred at the bottom of the tube and thus the surface of nanoparticles can be continuously renewed and exposed to the plasma for thin film deposition during the plasma polymerization process. A magnetic bar was used to stir the powders. The gases and monomers were introduced from the gas inlet during the plasma cleaning treatment or plasma polymerization. The system pressure was measured by a pressure gauge. A discharge by RF power of 13.56 MHz was used for the plasma film deposition.

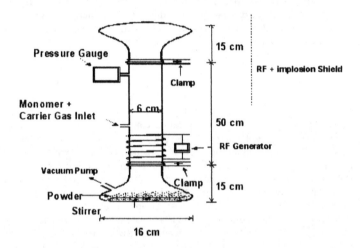

Figure 1. Schematic diagram of the plasma reactor for thin polymer film coating of the nanoparticles.

Before the plasma treatment, the basic pressure was pumped down to less than 2 Pa and then the plasma gases or monomer vapors were introduced into the reactor chamber. The operating pressure was adjusted by the gas/monomer mass flow rate. Pyrrole was used as the monomer for plasma polymerization. During the plasma polymerization process, the input power was 10 W and the system pressure was 25 Pa. The plasma treatment time was 240 min. Per batch 40 grams of powder were treated.

After the plasma treatment, the nanoparticles of alumina were examined by using high-resolution transmission electron microscopy (HRTEM), scanning electron microscopy (SEM), and x-ray diffraction (XRD). The high-resolution TEM experiments were performed on a JEM 4000EX TEM. The TOFSIMS analyses were performed on an ION-TOF Model IV instrument. Vickers microhardness testing was used to determine the hardness values of consolidated bulk samples.

The consolidation of the coated nanoparticles was carried out in a straightforward fashion. After coating, both coated and uncoated powders were pressed into pellets with a 13-mm diameter die. The applied pressure for each pressing was 19 MPa. The pellets were then heat-treated in a box furnace in air. The heat treatment temperatures used were 250 °C, 350 °C, 550 °C, and 800 °C. At each temperature the sample was held for various times: 60 min, 120 min, 240 min, and 360 min, and 480 min. After heat treatment, the pellets were air cooled to room temperature.

RESULTS AND DISCUSSION

Fig. 2 shows the bright field image of the coated nanoparticles. An ultrathin thin film of pyrrole can be clearly seen over nanoparticles of different sizes. The ultrathin film is marked by the double-lines in Fig. 2a and 2b. The thickness of ultrathin film is approximately 2 nm and appears to be uniform surrounding the particle surface. Particularly important, although these particles have different diameters, the film remains the same thickness indicating a uniform distribution of active radicals in the plasma chamber. Fig. 2b is the high-resolution image of a particle with a uniform nanocoating. The ultrathin film is tightly bound to the particles. The film was identified as typical amorphous structure by HREM observation over different particles.

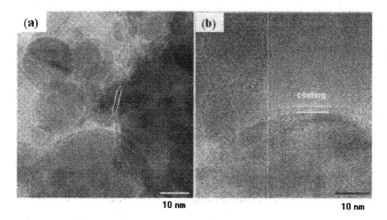

Figure 2. (a) Bright-field TEM image of the coated alumina nanoparticles (scale bar = 20 nm). The coated layers are marked by the double lines. (b) HRTEM image showing the amorphous, pyrrole-coated nanoparticle surfaces (scale bar = 10 nm).

As a comparison, the plasma coating of a different oxide nanoparticle ZnO is also shown here for demonstrating the coating uniformity. Figure 3a shows the low magnification of a ZnO particle coated with an acrylic acid (AA) film. As can be seen, the coating is uniform all the way around the entire surface of the particle. In our TEM observation we found that, although these particles have different diameters, the film remains the same thickness indicating a uniform distribution of active radicals in the plasma chamber. Figure 3b is an image at higher magnification. The uniformity of the AA thin film can be clearly seen in this photograph. The film thickness is about 5 nm.

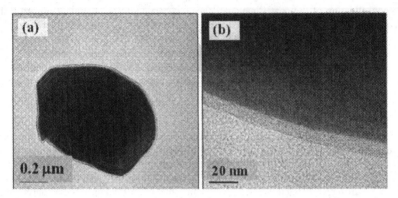

Figure 3. (a) Bright-field TEM image of the AA coated ZnO nanoparticles at low magnification. (b) HRTEM image showing the AA coated ZnO nanoparticle surfaces at higher magnification.

Fig. 4 shows the SEM micrographs of consolidated alumina with and without polymer coating. Both samples were heat treated at 250 °C for 60 min. As can be seen in Fig. 4a, the uncoated particles exhibit a highly porous structure as expected, since there is no sintering effect at such a low temperature. In contrast, under the same heat-treatment condition, the coated particles adhere to each other resulting in a dense matrix as shown in Fig. 4b. The density of the coated sample is estimated to be at least 95%, which is comparable to that of a high-temperature sintered counterpart.

As we heat-treated these samples at a much higher temperature, the microstructure showed rapid grain growth, however, in an inhomogeneous fashion. As exemplified in Fig. 5a, a sample heat-treated at 250 °C for 360 min. showed a highly densified matrix. Fig. 5b shows the surface of a sample that was heat-treated at 800 °C for 360 min. As can be seen, the grain growth is evident but only in some local regions. Due to this localized grain growth, internal stress is not evenly distributed. At this temperature we assume that the polymer coating has been entirely decomposed.

Figure 4. SEM micrographs showing the consolidated alumina (a) with and (b) without pyrrole coating. Both samples were heat treated at 250 °C for 60 min.

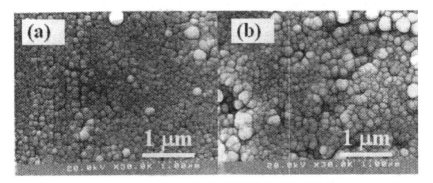

Figure 5. SEM micrographs showing (a) the sample heat-treated at 250 °C for 360 min. and (b) the sample heat-treated at 800 °C for 360 min. Both were pyrrole coated.

Although this research is mainly focused on the study of interface adhesion, we have carried out the heat-treatment at much higher temperatures such as 1000 °C for comparison purposes. At 1000 °C, the polymer film in the sample should completely decompose, however, the C-C bonds still remain between particles. These bonds can be responsible for considerable interface attraction among particles. The coated polymer film can also change the surface energy, which leads to a better dispersion of particles. Fig. 6 shows the SEM images of the sample that was heat-treated at 1000 °C for 12 hours. As can be seen in Fig. 6, the polymer coating, although entirely burned out at this temperature, has resulted in significant microstructural difference. We have found that the sample with the polymer coating (Fig. 6b) is considerably denser than that of non-coated one (Fig. 6a). The grain size of the uncoated sample is also noticeably larger than its coated counterpart.

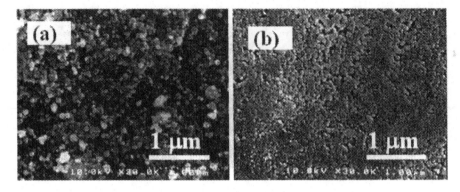

Figure 6. SEM micrographs showing the samples heat-treatment at 1000 °C for 12 hours for uncoated (a) and coated (b) nanoparticles.

Fig. 7a shows the microhardness values for the samples that were heat treated up to 800 °C for 60 min. The nanoparticles used in these samples before consolidation were coated with the thin films of pyrrole. The short heat-treatment time was to ensure that no sintering effect would result at elevated temperatures. As can be seen in this figure, there is no significant change in microhardness from the samples heat-treated in such a wide temperature range, suggesting a similar particle-to-particle bonding mechanism. However, above 350 °C, no adhesive polymer thin film should exist on nanoparticle surfaces, that may cause the gradual decrease in microhardness above this temperature. Fig. 7b shows that microhardness values versus heat-treatment time for consolidated alumina nanoparticles at heat-treatment temperatures indicated. As can be seen, the heat-treatment time does not seem to have significant effect on microhardness. This may also be an indication that the adhesive film is responsible for bonding in this range of temperature and time. The exact consolidation procedure was also applied to un-coated nanoparticles of alumina. However, their microhardness values were found to be about 30-time less than those shown in Fig. 7.

To study the interface adhesive behavior we have carried our HRTEM experiments on the consolidated nanoparticles. Figure 8 shows the HRTEM images of the pyrrole-coated alumina nanoparticles. In these images, the crystal Al_2O_3 lattice is quite apparent with the amorphous polymer film at the particle interfaces. It is also apparent that there is a layer of ultrathin film (NSA) between the nanoparticles, which is extremely thin, on the order of a few nanometers. The thin films on the nanoparticles exhibited roughened surfaces as shown in Fig. 8, which is in contrast to the smooth surfaces shown in Fig. 2. Such a deformed surface appears to be a result of particle-particle de-bonding during the breaking of the clusters for TEM experiment. The rough surfaces also indicate the adhesive characteristics of the interfacial bonding between the nanoparticles. In Fig. 9, we can see that the alumina nanoparticle interfaces appear to be "glued" together with pyrrole elastically deformed. Both edges of the joint also appear to be curved, indicating a quite flexible pyrrole film at the interfaces.

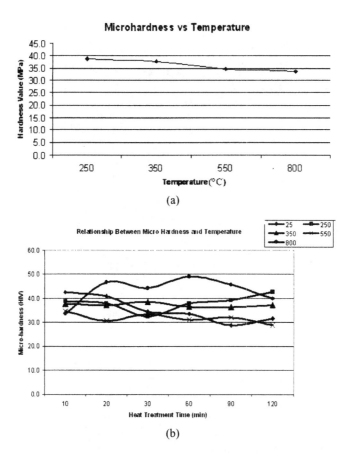

Figure 7. a) Microhardness vs. temperature for the samples that were heat-treated up to 800 °C for 60 min. and b) microhardness vs. heat-treatment time at the heat-treatment temperatures indicated. Note that the alumina nanoparticles in the samples shown in this figure were coated with pyrrole thin films before consolidation.

Two major mechanisms of adhesion that we may consider for coated nanoparticles are: mechanical interlocking and adsorption theory.[10-15] The theory of mechanical interlocking essentially proposes that mechanical keying, or interlocking, of the adhesive into the irregularities of the substrate surface is the major source of intrinsic adhesion. However, the attainment of good adhesion between smooth surfaces exposes this theory as not being of wide applicability. And, considering the role of mechanical interlocking, it is firstly debatable whether mechanical interlocking really does occur and secondly, even if it does, to assess its contribution to the strength and stability of the interface is difficult. However, based on this theory, mechanical interlocking occurs on a large-scale surface roughness of the order of several hundred microns. Hardly can it be used to explain the interfacial adhesion when the particle size decreases sharply to the order of nanometer.

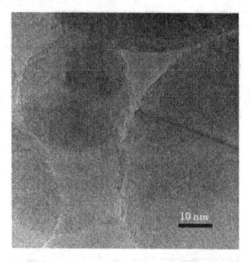

Figure 8. Bright-field TEM image of the coated alumina nanoparticles that have been consolidated at 250 °C for 360 min. (scale bar = 10 nm). The coated layer surfaces are roughened due to de-bonding indicating an adhesive behavior at the nanoscale.

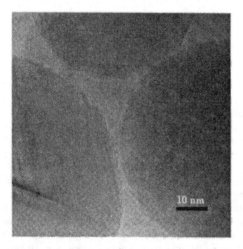

Figure 9. TEM image of the same sample shown in Fig. 8 with the elastically deformed adhesive layer at the particle interfaces (scale bar = 10 nm).

The molecular forces in the surface layers of the adhesive and substrate greatly influence the attainment of intimate molecular contact across the interface, called the adsorption. The attainment of interfacial contact is invariably a necessary first stage in the formation of strong and stable adhesive joints. The next stage is the generation of intrinsic adhesion forces across the interface, and the nature and magnitude of such forces are extremely important. They must be sufficiently strong and stable to ensure that the interface does not act as the "weak link" in the joint, either when the joint is initially made or throughout its subsequent service life.[9]

Other mechanisms include the diffusion and the electronic theories.[9,16-18] In the diffusion theory, the intrinsic adhesion of polymers is established through mutual diffusion of polymer molecules across the interface. This requires that the macromolecules, or chain segments of the polymers (adhesive and substrate) possess sufficient mobility and are mutually soluble. In a nanostructured medium, the coated particles are compacted by a three-dimensional network with multiple interfaces around an individual nanoparticle. The diffusion process is therefore fundamentally different from the planar interface adhesion.

The electronic theory treats the adhesive/substrate systems as a capacitor, which is charged due to the contact of the two different materials.[9] Separation on the parts of the capacitor, as during the interface rupture, leads to a separation of charge and to a potential difference, which increases until a discharge occurs. Adhesion is presumed to be due to the existence of these attractive forces across the electrical double layer. The adsorption theory states that the materials will adhere because of the interatomic and intermolecular forces, which are established between the atoms and molecules in the interfaces of the adhesive and substrate. Again, these theories apply to macroscopic interfaces and cannot easily be used to model the nanoparticle interfaces, although these mechanisms are assumed to exist.

Therefore, to study the fundamental mechanism of adhesion between nanoparticles, new theoretical modeling work is required for a varied nanosurface structure. As such, a mathematical model has to be developed to study particle-particle adhesion instead of a flat surface macroscopic adhesion. Our current work is focused on the modeling work based on the results of the tensile test.

SUMMARY

In summary, we have deposited an ultrathin film of pyrrole on the surface of alumina nanoparticles by means of a plasma polymerization treatment. The polymer layer is not only uniform on all particle sizes, but also extremely thin with a thickness of 2 nm. Such ultrathin film deposition characteristics are essential in establishing multi-layer nanostructures, particularly for adhesive bonding at nanoparticle interfaces. With such a thin polymer film on nanoparticles we have been able to observe, by means of HRTEM, a unique adhesive behavior at the nanoparticle interfaces. Based on this adhesive film we have consolidated the nanoparticles at temperatures much lower than sintering temperature. Although the operating adhesion mechanism requires further studies we believe, based on the experimental data from this study, that intrinsic adhesion forces are responsible for the observed interfacial bonding. Our future work will focus on the interface study in terms of structure, adhesion behavior, and related mechanical properties. Improved adhesive coatings will also be selected for coating of the nanoparticles.

ACKNOWLEDGEMENTS
The TEM analyses were conducted at the Electron Microbeam Analysis Laboratory at the University of Michigan, Ann Arbor, Michigan. This research was supported in part by a grant from NSF, DMII division, No. DMI-9713715.

REFERENCES
[1] J. S. Reed, "Principles of Ceramics Processing," 2nd ed. (John Wiley & Sons, Inc, New York), 1988.

[2] J. E. Buek and J. H. Rosolowski, "Treatise on Solid State Chemistry," vol. 4, N. B. Hannay, ed., (Plenum, New York), 1976.

[3] J. P. Singh, D. Shi, and D. W. Capone, "Mechanical and superconducting properties of sintered composite $YBa_2Cu_3O_{7-x}$ delta tape on a silver substrate," Appl. Phys. Lett., 53, 239-241 (1987).

[4] D. Shi, D. W. Capone II, G. T. Goudey, J. P. Singh, N. J. Zaluzec, and K. C. Goretta, "Sintering of $YBa_2Cu_3O_{7-x}$ compacts," Mater. Lett., 6, 217-221 (1988).

[5] J. P. Singh, H. J. Leu, R. B. Poeppel, E. Van Voorhees, G. T. Goudey, K. Winsley, and D. Shi, "Effect of Silver and Silver oxide Additions on the mechanical and Superconducting Properties of $YBa_2Cu_3O_{7-x}$ Superconductors," J. Appl. Phys., 66, 3154-3159 (1989).

[6] K. C. Goretta, O. D. Lacy, U. Balachandran, D. Shi, and J. L. Routbort, "$YBa_2Cu_3O_x$ toughened by ZrO_2 additions," J. Mater. Sci. Lett., 9, 380-381 (1990).

[7] S. Eufinger, W. J. van Ooij, and T. H. Ridgway, "DC plasma-polymerization of pyrrole: Comparison of films formed on anode and cathode," Journal of Appl. Pol. Sci., 61, 1503 (1996).

[8] W. J. van Ooij, S. Eufinger, and T. H. Ridgway, "Surface Modification of Textile Fibers and Cords by Plasma Polymerization," Plasma and Polymers, 1, 231 (1996).

[9] A.J. Kinloch, "Adhesion and Adhesives," Chapman and Hall, London, 1987.

[10] R. W. Siegel, Nanostructured Materials, 3, 1 (1993).

[11] J. R. Hunstberger, in Treatise on Adhesion and Adhesives, vol. 1, ed. R. L. Patrick (1967), Marcel Dekker, New York.

[12] J. T. Dickson, L. C. Hensen, S. Lee, L. Scudiero, and S. C. Langford, "Fracto-emission and electrical transients due to interfacial failure," J. Adhesion Sci. Technol., 8, 1285 (1994).

[13] R. G. Horn, and D. T. Smith, and A. Grabbe, "Contact electrification induced by monolayer modification of a surface and relation to acid–base interactions," Nature, 366, 442 (1993).

[14] R. G. Horn and D. T. Smith, "Contact Electrification and Adhesion Between Dissimilar Materials," Science, 256, 362 (1992).

[15] J. N. Israelachvili, "Thin film studies using multiple-beam interferometry," J. Coll. Interface Sci., 44, 259 (1973).

[16] F. M. Fowkes, "Attractive Forces at Interfaces," Ind. Eng. Chem., 56, 40 (1964).

[17] A. J. Kinloch, W. A. Dukes, and R. A. Gledhill, in Adhesion Science and Technology, ed. L. H. Lee, (1975) Plenum Press, New York.

[18] E. P. Plueddemann, Silane Coupling Agents, (1982), Plenum Press, New York.

PLASTIC DENSIFICATION AND GRAIN GROWTH OF NANOCRYSTALLINE ZIRCONIA POWDERS

Chiraporn Auechalitanukul, Michael J. Roddy[1] and W. Roger Cannon
Rutgers, The State University of New Jersey
Ceramic and Materials Engineering Department
607 Taylor Road, Piscataway, NJ 08854

ABSTRACT

Zirconia with 2.5% Y_2O_3 was sinter-forged using two different methods to prepare the predensified (70%) specimens used for forging. Method 1 presintered the commercial TZ-2.5 TOSOH cylindrical specimens at 500°C. Specimens were then forged to full density at 1200°C and forging stresses >30 MPa. Fully dense specimens had a 165 nm grain size. Method 2 plasma sprayed the commercial TZ-2.5Y TOSOH powder directly into water to quench in a nanostructure in the particles. Predensified specimens (70%) were fabricated by hot pressing at 1250° C then sinter-forging at 1200°C. Densification is by creep. Specimens densified to 75% of theoretical at 30 MPa but fractured at higher stresses. Slower densification by method 2 resulted primarily because of a lower driving force and larger grain size.

INTRODUCTION

Several studies have shown that sinter-forging, the pressure-assisted sintering of unconstrained compacts subjected to uniaxial pressure, is very effective in densifying ceramics. This study used sinter-forging to densify ZrO_2 compacts at low temperatures and short times to avoid grain growth in an attempt to achieve a nanocrystalline structure. Many studies have shown that SZP (stabilized zirconia polycrystals) with 3% Y_2O_3 can be fully pressureless sintered at 1350-1500°C with a grain size of 400 nm. Subsequently, SZP can be superplastically formed at rates of ~10^{-4} sec^{-1}.[1] Reducing the grain size to the nanocrystalline size, i.e. <100 nm may allow SZP to be formed about >4^2= 16 times faster, which may make the process more viable. In this paper, we consider methods of forging nanocrystalline SZP.

Nanocrystalline ceramics are normally fabricated directly from nano-sized particles; however, they are often not cost effective. Nano-sized powders adsorb more moisture, and may not be as processing friendly as micron-sized powders. In this paper, we compare two methods of sinter-forging: method 1 uses the nano-sized powder and method 2 uses micron-sized powder (5-50 μm) containing nanocrystalline grains within it. Method 2 relies on plastic deformation of the particles by creep to densify. However, since they both start from nano-sized grains, the diffusion distances are similar and so the kinetic terms, A and A', in the equations for densification rate, $\dot{\varepsilon}_v$, and creep rate, $\dot{\varepsilon}_c$, should be similar.[2]

$$\dot{\varepsilon}_v = A\left(\frac{\sigma_z}{3} + \Sigma\right) \tag{1}$$

$$\dot{\varepsilon}_c = A'\sigma \tag{2}$$

---

[1] Now at Procter and Gamble Co, Este Process Technology Center, Cincinnati Ohio

where Σ is the sintering stress and σ is the applied axial stress. Method 1 has a higher driving force since Σ for a nanocrystalline ceramic is not trivial, and so method 1 has a kinetic advantage over method 2 depending on the applied stress.

The micron-sized powder with nanocrystalline grains was fabricated by plasma spraying spray-dried powder into water to quench in the nanostructure. It is recognized that since the spray-dried particulates are melted at high temperature before rapid cooling, the non-equilibrium phase tetragonal polymorph t'- ZrO_2 will be formed using method 2. As the t'-ZrO_2 is non-transformable, the transformation toughening may not occur. [3] However, the strength may still be very high.

EXPERIMENTAL PROCEDURE
Method 1 : Sinter-forging nano-sized particles
The commercial 2.5 mole percent yttria stabilized zirconia powder (TZ-2.5, TOSOH Corporation, Tokyo, Japan) was used as the starting powder. It was produced via a hydrolysis and calcination, and then spray dried into the uniform spherical agglomerates. The spray-dried powder was loaded into a double-action uniaxial die and dry pressed at low compaction pressures (< 7 MPa) before cold isostatically pressing at 276 MPa. To remove the binder introduced during spray drying and give the specimen some strength, the samples were pre-sintered at 500 °C for 48 hours with a heating rate of 60 °C/hour, and then cooled to room temperature at the same rate. A density of 70 % theoretical was obtained after pre-sintering.

In order to determine the densification rate during forging, individual specimens were used for each sintering time. The density and grain size were measured after each time. Sinter-forging was at 1200 °C and stresses between 4 and 60 MPa were applied for 2-60 minutes.

For grain size measurement the x-ray Scherer equation was used. The results compared well with measurements made from transmission electron microscopy (TEM) analysis. X-ray diffractometer (SIEMEN D500) was operated at 40 kV and 30 mA, and TEM (TOPCON EM0028) was operated with 200 kV accelerating voltage. For TEM analysis, samples were mechanically thinned and ion milled to be electron transparent.

Method 2 : Sinter-forging nanocrystalline micron-sized particles.
Spray dried powder from the same bottle of TZ-2.5 TOSOH zirconia, used for method 1, was plasma sprayed (Argon working gas and arc power : 600 A, 75 V) and water-quenched (stand off distance of 30 cm.) The plasma-sprayed powder was loaded into the 5 mm-diameter alumina cylindrical die and then compacted under a uniaxial pressure of 25 MPa before hot-pressing in air in the same alumina die under a pressure of 65 MPa for 40 minutes at 1250 °C. These samples were then sinter-forged. The predensified samples had the same density (70% theoretical density) as the specimens in method 1.

Unlike method 1, axial strain was measured using a linear variable displacement transducer (LVDT) and radial strain using a laser extensometer (Laser Mike), both axial and radial deformation strains were collected simultaneously. Consequently, the density of sinter-forged sample prepared from plasma-sprayed powder was determined instantly and so only one hot-pressed sample was used for each pressure in order to plot density vs. time. Unfortunately, the laser extensometer was not sensitive enough to measure the small strain rates when <15 MPa was applied at 1200°C and specimens fractured if >30 MPa was used.

X-ray diffractometer was utilized to estimate the grain size which was compared to the result obtained from TEM analysis. Since sample has low density, crushing the sample was used to prepare the TEM sample instead of mechanically thinning.

RESULTS AND DISCUSSION

Method 1: Sinter-forging nano-sized powder

For method 1, the density of sinter-forged samples as a function of time at 1200 °C under different applied pressure of 0, 4, 8, 15, 30, and 60 MPa is shown in the Fig. 1. Full density was only achieved at the highest stresses. At 60 MPa, a density of 99.9% theoretical density was achieved within 10 minutes.

Results of grain size vs. time are plotted in Fig. 2. Note that the grain growth rate increases as applied pressure increases indicating dynamic grain growth.

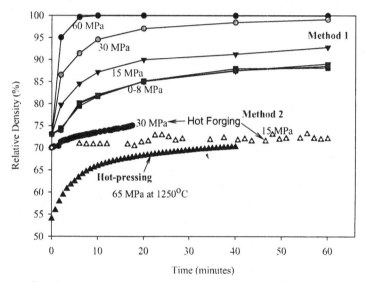

Figure 1. Density vs. time at 1200° C of sample prepared by both method 1 and method 2 using 2.5 mole percent yttria stabilized zirconia.

Method 2: Sinter-forging nanocrystalline-containing micron-sized powder

Fig. 1 compares densification rates obtained by method 1 with those obtained by method 2. Also shown are the results of hot pressing. Hot pressing the plasma sprayed powder of method 2 inside a die increased the density from the initial 54% to 70% at 1250° C. Note that the final rate of hot pressing densification is about the same at 1250°C and 65 MPa as the forging rate is at 1200°C and 30 MPa. The slower rate of hot pressing is probably due to die wall friction. Comparing the two methods, the densification by method 1 is many times faster than method 2. Therefore, very little density has been achieved by method 2, whereas full densification is achieved by method 1.

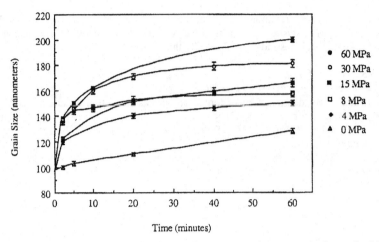

Figure 2. Grains size vs. time at 1200 °C under different applied pressure by method 1.

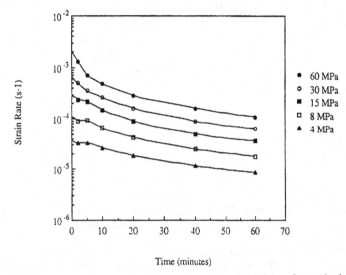

Figure 3. Creep rates calculated from Eq. (3) as a function of time for method 1.

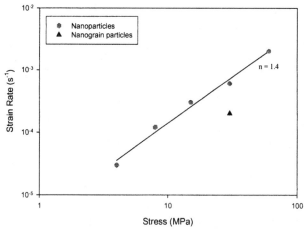

Figure 4. Comparison of initial creep rates calculated from Eq. (3) between method 1 and method 2.

There are several possible reasons for the comparatively low densification rates by method 1. To a first approximation, the densification rate is the product of a kinetic term and the driving force as per Eqs. (1) and (2). For method 1, the driving force consists of the applied stress and the sintering stress (surface energy driving force written in terms of stress). The sintering stress is derived from surface area reduction during sintering. The sintering stress for the intermediate sintering stage may be estimated from the equation $\frac{\gamma_{ZrO_2}}{r}$ where γ_{ZrO_2} is the surface energy and r is the pore radius.[4] Assuming γ_{ZrO_2} to be ~0.8 J/m^2 [5] and since the pore radius is about ~30 nm [6], then the sintering stress is 26 MPa which is larger than $\frac{\sigma}{3}$ = 10 MPa (σ = 30 MPa).

The kinetic terms in Eqs. (1) and (2) were compared by determining the effective creep strain rates during forging by method 1 and 2. The strain during densification using method 1 can divided into the dilatational strain and the deviatoric strain or creep. Assuming that densification is isotropic and creep is constant volume, the deviatoric or creep strain is given by [7]

$$\varepsilon_c = \frac{2}{3}(\varepsilon_z - \varepsilon_r) \qquad (3)$$

where ε_c, ε_z, and ε_r are a creep strain, axial strain and radial strain, respectively. From Eq. (3), the creep strain is calculated and subsequently the creep strain rate. In order to compare creep rate as a function of stress, the microstructure must be constant. Because of dynamic grain growth, grain size depends both on stress and time. Thus the grain size during steady state creep will not be constant at a given time, independent of stress. Thus in Fig. 3 strain rates were plotted as a function of time and extrapolated back to zero time where all specimens had the same grain size, and same density (70% dense). Fig. 4 compares creep rates from method 1 and the single point available for method 2. (The data for 15 MPa contained too much scatter to be considered.)

The creep rate of the sample produced by method 1 is faster than the creep rate from method 2 by a factor of about 2. Results were also compared with literature values for creep. Jimenez-Melendo et al. [1] found that the creep rates of Y-SZP of various Y_2O_3 content could be made to fall on the same straight line if compensated for grain size, d, by plotting $\dot{\varepsilon}d^2$ vs. σ. Fig. 5 uses this method to compare results from method 1 corrected for porosity [8] with literature values [1] of creep rates at 1200°C taken from their paper.(Results were for 1250°C creep but used Q_c=460 kJ/mole to correct for temperature.) Agreement with the literature creep rates is good though the stress dependence of method 1 is lower. Method 2 results are not placed on the graph because of the uncertainty of the grain size as discussed below. If $\dot{\varepsilon}d^2$ for the single point from method 2 were made to coincide with the literature value of creep in Fig. 5, the grain size would need to be 200 nm. If it were made to coincide with method 1, it would be 160 nm.

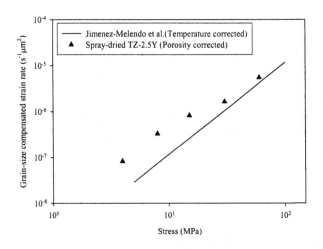

Figure 5. Creep rate compensated for grain size, $\dot{\varepsilon}d^2$ vs. stress, σ. Comparing literature values with method 1 determined by Eq. (2).

The apparent grain size of the particles in method 2 determined from X-ray using the Scherer equation were 45 nm initially and 53 nm at the end of the test. The literature, however, reports that t'- ZrO_2 quenched from the melt contains small domains bounded by stacking faults.[9] Thus the x-ray domain size is likely not the true grain size, i.e. the grain size that controls creep. It was difficult to determine the grain size of method 2 specimens by microscopy since they were only 75% dense. Fig. 6 shows a TEM micrograph of a crushed portion of the hot-pressed specimen. Most grains, indeed are on the order of 50 nm but >100 nm grains were also observed. The more accurate estimate might be the one coming from the kinetic argument above.

Method 2 samples do not fully densify both because they lacks the driving force from surface energy and because the grain size was larger. If higher pressing pressures were used the driving force from surface energy would not be so significant. Several other factors are important. First, because of the large particle size in method 2, the pre-densified, 70%, dense specimens used for sinter-forging were not as strong as the pre-sintered specimens of method 1 and so they tended to crack around the edges even at 30 MPa. Secondly, stress is concentrated at the contact points between the fine particles (method 1). Although there is also a contact stress between the large particles (method 2), this stress is felt by only a few grains near the contact. Third, grain boundaries formed from the melt are generally lower energy and have fewer impurities in them. Therefore, they may neither be as efficient sources or sinks of vacancies nor will diffusion be as fast in the grain boundary. Finally, diffusion rates in t' phase may be different than in t phase. Data is not available to consider at this point.

Although method 2 was not successful here, it has been successful in making high-density specimens of some two phase materials and is actively pursued in our laboratory. If the predensifying hot pressing step could achieve higher densities then higher pressures could be used during forging and full density may be obtained.

Figure 6. TEM micrograph of crushed specimen after hot pressing.

CONCLUSIONS

Two alternative of methods of sinter-forging were used to fabricate 2.5Y-SZP. The first method started with commercial TZ-2.5 TOSOH powder presintering to 70% dense and then forged at various stresses. Full density could be achieved in less than 1 hour for stresses >30 MPa. The grain size was 165 nm. The second method started with water-quenched plasma sprayed (same composition) powder and attempted to densify it by sinter-forging. It was assumed that the grain size of the water-quenched powder was in the nano-size range and that densification by plastic deformation (creep) might be just as fast. It was found, however, that the forging rates of the water quenched powder (method 2) were much slower than the presintered powder (method 1). Several reasons were given including driving force and grain size.

REFERENCES

[1]M. Jimenez-Melendo, A. Dominguez-Rodriguez, and A. Bravo-Leon, "Superplastic Flow of Fine-Grained Yttria-Stabilized Zirconia Polycrystals: Constitutive Equation and Deformation Mechanisms," *Journal of the American Ceramic Society*, **81** [11] 2761-2776 (1998).

[2]J. W. Evans and L. C. De Jonghe, "Sintering of Powder Compacts"; pp.410-447 in *The Production of Inorganic Materials*, McMillan, New York, 1991.

[3]R. A Miller, J. L Smialek, and R. G. Garlick, "Phase Stability in Plasma-sprayed Partially Stabilized Zirconia-Yttria," pp. 241-253, in *Advances in Ceramics*, vol. 3 Edited by A. H. Heuer and L. W. Hobbs., The American Ceramic Society, Inc., Columbus, Ohio, 1981.

[4]R. C. Garvie, "The Occurrence of Metastable Tetragonal Zirconia as a Crystallite Size Effect," *Journal of Physical Chemistry*, **69** [4] 1238-12343 (1965).

[5]G. Skandan, H. Hahn, B. H. Kear, M. Roddy, and W. R. Cannon, "Processing of nanostructured zirconia ceramics"; pp. 207-212 in *Materials Research Society Symposia Proceedings*, vol. **351**, 1994.

[6]A. Allen, S. Krueger, G. Skandan, H. Long , H. Hahn, J. C. Parker, and M. Parker, "Microstructural Evolution during the Sintering of Nanostructured Ceramic Oxides," *Journal of the American Ceramic Society*, **79** [5] 1201-1212 (1996).

[7]R. Raj, "Separation of Cavitation Strain and Strain During Deformation," *Journal of the American Ceramic Society*, **65** [3] C-46-49 (1982).

[8]T. G. Langdon, "Dependence of Creep Rate on Porosity," *Journal of the American Ceramic Society*, **55** [12] 630-634 (2004).

[9]V. Lanteri, A. H. Heuer, and T. E. Mitchell, "Tetragonal Phase in the System ZrO_2-Y_2O_3," pp.118-130, in *Advances in Ceramics*, vol. 12, Edited by N. Clausson, M. Ruhle, and A. H. Heuer, The American Ceramic Society, Columbus, Ohio, 1983.

FUNDAMENTAL RHEOLOGICAL MODELING TECHNIQUE AND FRACTURE MECHANICS PRINCIPLES OF DIAMOND-CONTAINING NANOCOMPOSITES

Maksim V. Kireitseu
Composite Nano/Materials Research Center and Laboratory
University of New Orleans
New Orleans, LA 70148-2220, USA
E-mail: m.kireitseu@tridentworld.com

ABSTRACT

The principal goal of the paper is to examine fracture mechanics and rheological behavior of multi-layered nanostructured materials and composites using a rheological modeling. The core principal of Newton's fundamentals was combined with rheological modeling for describing correlation between stress and deformation of nanostructured materials. Alumina-based and diamonds - containing composites (diamonds nanoparticles - aluminum, diamonds nanoparticles - polymer matrix) were studied by indentation at micro- and nano-scales. Rheological modeling as a method for determining the fracture of nanocomposites is described. Analysis of the models and experimental results revealed better understanding of nanocomposites failure and degradation mechanics. Only a deformation or a fracture load should be measured for this modeling. By measuring this minimum load, an accurate estimate of fracture may easily be calculated by applying this rheological modeling.

INTRODUCTION

Recently, Lawn et al.[1] revealed that there are two fundamentally different fracture/damage mechanisms in ceramics that are usually determined by their microstructures. First class of very brittle materials consists of single crystals, fine-grained ceramics or the like. In this ceramic, localized loading produces macroscopic cone cracks when the indentation load exceeds a certain critical value for a given material.[2-6] This type of cone cracking may result in severe strength degradation.[7, 8]

In some of new ceramic materials with advanced strength behavior, such as aluminum oxide materials, the cone cracking is suppressed because of crack deflection along the sub materials, its micro granular submerged particles and hard micro structural elements. Along with some cone and ring cracks a damage zone is formed within the region of high shear-compression concentration just under the indenter.[8-11] In some cases both the highly destructive macroscopic cone cracking and the damage dispersion occur in the material. However, the growth and linkup of the micro cracks under a high load or repeated loading conditions lead to severe surface damage and ultimate material peeling off from the surface.[12, 13]

One potential approach for avoiding such a fracture is to use a multi-layered structure, in which materials strengthen each another. The view to this approach comes from the characteristic of the spatial distribution of Hertzian shear stresses in the aluminum oxide material. Therefore, if the micro structural weaknesses, such as pores and defects, can be strengthen along the r-z plane, the failure and cracking of material, and its dispersion and failure can be reduced to a certain degree.

RHEOLOGICAL MODELLING AND FRACTURE MECHANICS

Several rheological models that were applied for resins, nanostructured conglomerates, viscous-elastic reinforced materials to investigate mechanical properties of a composite system.[11-14] However, despite of some advantages the known models are not applied to composite systems and are not experimentally confirmed. Based on the past findings the number of requirements to rheological modeling of nanocomposite systems is proposed.

The following requirements may be effectively used in rheological modeling of nanosystems: a) Since the structure of the system includes hard aluminum oxide layer and the plastically deformed substrates, irreversible deformations have to be considered as plastic deformations. Plastic materials deform only when loading exceeds ultimate yield strength; b) If loading is less than the yield strength, deformation increases step-by-step to a final value at constant load; c) Plastic deformations are added under cyclic loading; d) Function of deformation vs. time has a linear segment on a plotted curve at constant loading; e) Retardation of deformations (elastic return) can be observed during unloading; g) Stresses are relaxed at constant loading rate.

In addition to the above stated conditions, the following mathematical requirements must be considered: (a) rheological equations must have a final decision (order of a required differential equation of stress and deformation should not exceed numbers of possible conditions of physical limitations); (b) equations and formulated problems must be solved concerning stress or strain rates. A rational choice of an adequate model may be made by comparison of calculated and experimental results.

It is expected that aluminum matrix reinforced by diamond nanoparticles can be presented by an "elastic-viscous-plastic" rheological model (model 1) effectively shown for asphalts and polymer matrix composites.[14] A mechanical prototype of the model is shown in Fig. 2a. On the other side, polymer matrix reinforced by diamond nanoparticles can be presented by a simple rheological model consisting of two elastic elements and a viscous element (model 2 shown in Fig. 2b). A mechanical prototype of the model 2 can be presented as a Maxwell's model and an elastic element connected. A structural equation of the sequentially joined models of the system can be written in another form: (H ‖ N ‖ St-V) - (H-N ‖ H).[14] The type of rheological equation usually depends on load rating and shape of stress applied to the models.

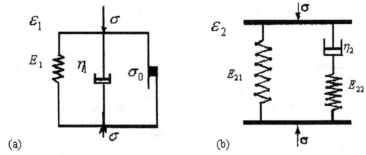

Fig. 2. Rheological models of composite systems

Deformation of the sequentially joined models is calculated as follows:

$$\varepsilon = \varepsilon_1 + \varepsilon_2, \tag{1}$$

If the loaded system has the stress condition $\sigma > \sigma_0$, then the equation of the composite system (model 1 in Fig. 2a) can be written as:

$$\sigma = \sigma_0 + E_1\varepsilon_1 + \eta_1\frac{d\varepsilon_1}{dt} \tag{2}$$

where σ_0 σ, are stresses at initial and final time of the loading respectively; E_1 is the modulus of elasticity of the layer; η is the coefficient of viscosity; ε is the deformation.

For the second part of the system (model 2 in Fig. 2b) including elastic-viscous polymer layers, the differential equation is defined by:

$$\frac{d\sigma}{dt} + \frac{E_{22}}{\eta_2}\sigma = (E_{21} + E_{22})\frac{d\varepsilon_2}{dt} + \frac{E_{21}E_{22}}{\eta_2}\varepsilon_2 \tag{3}$$

where E_0, E_1 are moduli of elasticity of the material; η is the coefficient of viscosity.

The rheological equation of the sequentially connected models at the random loading conditions is given by:

$$\frac{\eta_1}{E_{22}}\frac{d^2\sigma}{dt^2} + \left(\frac{E_1}{E_{22}} + \frac{\eta_1}{\eta_2} + \beta\right)\frac{d\sigma}{dt} + \frac{E_1 + E_{21}}{\eta_2}\sigma - \frac{E_{21}}{\eta_2}\sigma_0 = \beta\eta_1\frac{d^2\varepsilon}{dt^2} + \left(\frac{\eta_1}{\eta_2}E_{21} + \beta E_1\right)\frac{d\varepsilon}{dt} + \frac{E_1E_{21}}{\eta_2}\varepsilon \tag{4}$$

By mathematical transformations the equation (4) can be written as:

$$\sigma(t) = e^{-\frac{A}{2}\Delta t}\left[sh\left(\frac{AD}{2}\Delta t\right) \cdot \frac{x-y}{D} + xch\frac{AD}{2}\Delta t\right] + \sigma_0 - x \tag{5}$$

where the coefficients of equation (5) can be found as follows:

$$x - y = \sigma_0 - \frac{S_0^{\cdot}}{B} + \frac{AS_2^{\cdot} - 2S_2^{\cdot}}{B^2} - \frac{2A^2S_2^{\cdot}}{B^3} - \frac{2\sigma_0}{A} + \frac{2S_1^{\cdot}}{AB}; \quad A = \frac{E_{22}}{\eta_1}\left(\frac{E_1}{E_{22}} + \frac{\eta_1}{\eta_2} + \beta\right); \quad B = \frac{(E_1 + E_{22})/E_{22}}{\eta_1\eta_2};$$

$$D = \sqrt{1 - 4\frac{\eta_1}{\eta_2} \cdot \frac{(E_1 + E_{21})/E_{22}}{(E_1/E_{22} + \eta_1/\eta_2 + \beta)^2}};$$

$$S_{21} = \frac{E_1 E_{21} E_{22}}{\eta_1 \eta_2} \cdot \frac{1}{2} k_2 \; ;$$

$$S_0 = \frac{E_{21} E_{22}}{\eta_1 \eta_2} \sigma_0 + \beta E_{22} k_2 + \frac{E_{22}}{\eta_1}\left(\frac{\eta_1}{\eta_2} E_{21} + \beta \cdot E_1\right) k_1 \; ;$$

$$S_1 = \frac{E_{22}}{\eta_1 \eta_2}\left(\frac{\eta_1}{\eta_2} E_{21} + \beta \cdot E_1\right) k_2 + \frac{E_1 E_{21} E_{22}}{\eta_1 \eta_2} \varepsilon_0$$

Equations 1 to 5 describe the rheological behavior of the composite systems reinforced by diamond nanoparticles. To understand the rheological behavior of the composite systems we will consider some major strain-stress conditions that are listed in the table I.

Table I. Rheological behavior and stress-deformation lows

№	Stress-Deformation conditions	Stress- Deformation Low	
		Deformation (ε) of composite system at stress σ =const	
1.	If $\sigma \leq \sigma_0$ and $\varepsilon_1 = 0$	$\sigma(t) = (E_{21} + E_{22})\varepsilon_2(t) - \frac{E_{22}}{t_{p2}}\int_0^t \varepsilon_2(\tau)e^{-(t-\tau)/t_{p2}}d\tau + Ce^{-t/t_{p2}}$ (7)	$\varepsilon(t) = \frac{\sigma}{E_{21}}\left(1 - \frac{1}{\beta}e^{-t/t_{p2}}\right)$ (6)
2.	If $\sigma > \sigma_0$ and $\varepsilon_1(0)= 0$	$\varepsilon(t) = \frac{\sigma - \sigma_0}{E_1}(1 - e^{-t/t_{p1}}) + \frac{\sigma}{E_{21}}\left(1 - \frac{1}{\beta}e^{-t/t_{p2}}\right)$ (8)	
		Unloading	
3.	If $\sigma_1 \leq \sigma_0$ the deformation of model 1 and model 2 during time t_1	$\varepsilon_1(t_1) = 0$ and $\varepsilon_2(t_1) = \frac{\sigma_1}{E_{21}}\left(1 - \frac{1}{\beta}e^{-t_1/t_{p1}}\right)$ (9)	
4.	The unloading process at the initial condition	$\varepsilon_2(t) = \varepsilon_2^*(t_1)e^{-(t-t_1)/t_{p2}}$ (10)	
5.	If $\sigma > \sigma_0$	$\varepsilon(t) = \frac{\sigma_1 - \sigma_0}{E_1}\left(1 - e^{-t_1/t_{p1}}\right)e^{-(t-t_1)/t_{p1}} + \frac{\sigma_1}{\beta E_{21}}\left(1 - e^{-t_1/t_{p2}}\right)e^{-(t-t_1)/t_{p2}}$ (11)	
		Loading at ε=const	
6.	If $\varepsilon_0(E_{21} + E_{22}) \leq \sigma_0$, then if $\varepsilon_1 = 0$ and $\varepsilon_2 = \varepsilon_0$ =const	$\sigma(t) = \varepsilon_0 E_{22}\left(\frac{E_{21}}{E_{22}} + e^{-t/t_p}\right)$ (12)	
7.	If $\varepsilon_0(E_{21} + E_{22}) > \sigma_0$ and $\varepsilon_1 = 0$, $\varepsilon_2 = \varepsilon_0$ =const	$\sigma(t) = C_{11}e^{-(\varphi+D)t/2} + C_{12}e^{-(\varphi-D)t/2} + S_{/\psi}$ (13)	

Explanations to the equations are given here. In equation 6 we determine that $t_{p2} = \beta\eta_2/E_{21}$, and $\beta = 1 + E_{21}/E_{22}$. Under an infinitely long loading, the ultimate rate of deformation is modified to σ/E_{21}. In the equation (6) the integration constant C is defined by the initial conditions of stress-deformation state of the composite system. If $\sigma(0)=0$ and $\varepsilon_2(0)=0$, then it follows that $C=0$.

At an infinitely long loading we obtain: $\sigma(t \to \infty) = \varepsilon_0 E_{21}$.

In the equation 13, we determine variables:

$$\varphi = \frac{E_{22}}{\eta_1}\left(\frac{E_1}{E_{22}} + \frac{\eta_1}{\eta_2} + \beta\right), \qquad \Psi = \frac{E_{22}}{\eta_1\eta_2}(E_1 + E_{21}), \qquad S = \frac{\sigma_0 + \varepsilon_0 E_1}{1 + E_1/E_{21}}, \qquad D = \sqrt{\varphi^2 - 4\Psi},$$

$$C_{11} = \frac{R_2 - R_1(\varphi - D)}{2D}, \qquad C_{12} = \frac{R_1(\varphi + D) - R_2}{2D}.$$

By applying the rheological models to the composite systems we can reveal the stress-deformation mode of the system that correlates with rheological and mechanical properties (plasticity, elasticity and viscosity), microstructure and construction of the systems. Deformation and stress modes are shown in Fig. 3 at sinusoidal-shape loading conditions.

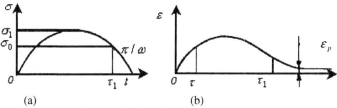

(a)   (b)

Fig. 3. Sinusoidal-shape loading (a), Deformation of the viscous-elastic-plastic models (b)

INDENTATION TECHNIQUE

Many authors[1-13] observed that deformation at certain stress is the function of an indentation track diameter (a) and a diameter of indenter (D). Therefore, it can be written as:

$$\varepsilon(t) = kd(t)/D \qquad (14)$$

The deformation rate is defined by:

$$\varepsilon(t) = \int_0^t \dot{\varepsilon}(s)\,ds = \varepsilon_0 + \dot{\varepsilon}_0 t - \frac{k_1 t^2}{2} \qquad (15)$$

The function of stress-deformation mode of sequentially joined models (Fig. 2) is described by Eq. (5). The function of stress-deformation mode of the composite system in Fig. 2a is defined by Eq. (2) with accepting stress-strain conditions at the linear correlation between deformation and time of loading. The function of stress-deformation mode of the model (Fig. 2b) is defined by:

$$\sigma(t) = E_{21}\left[\varepsilon(t) - \varepsilon_0\right] + \eta_2 \dot{\varepsilon}(t) - \frac{3k\eta_2}{2t_1} e^{-t/\tau} + \frac{k\eta_2^2}{E_{22}t_1^2}\left(1 - e^{-t/\tau}\right) \tag{16}$$

where $\tau = \eta_2/E_{22}$, $t_1 = R/V$; $k = 4/(3\pi\cdot(1-\mu^2))$.

The difference of ε and ε_0 in the brackets describes the deformations from loading. Of interest are the total deformation that equals the sum of the initial deformation of the trough surface layer and the deformation resulting from loading.

The above equations define the functions of stress-deformation mode expressing rheological behavior of the composite systems. Under some conditions of loading simplified rheological models (Fig.2) can provide adequate results that will be then well fitted to the experimental base. The function of contact stress rate is modified to:

$$P(t) = \sigma(t)\left[\pi R V t + \frac{\pi d^2(t)}{4}\right] \tag{17}$$

where R is the indenter radius, V is the velocity of indenter loading, t is the time of loading, $d(t)$ is the function of indentation track depth vs. applied load, $\sigma(t)$ is the function of stress rating at applied load.

The function Eq. 16 uses E_{21}, E_{22}, η_2 parameters that characterize mechanical properties of polymer layers. E_{21} and the coefficient η_2/E_{22} can be calculated from experiments on the stress relaxation at loading/unloading cycles.

Time of strain relaxation (τ) and Young's modulus E_{22} can be defined by investigating a reaction of elastic components of the composite material on applied load when time is approximated to zero (t→0). Under loading the composite material will be strained, but its elastic components of the composite material composite materials will be deformed first. Therefore, accepting above condition it follows at t→0:

$$\frac{d\sigma}{dt}\mathbf{I}_{t\to 0} = \dot{\varepsilon}_0(E_{21} + E_{22}) \tag{18}$$

Equation (18) defines relation between Young's moduli of the model (Fig. 1b). Also Young's modulus of the model 2 may be written as $E = E_{21} + E_{22}$. On the other hand, contact Young modulus may be calculated by:

$$\frac{dP}{dt}\mathbf{I}_{t\to 0} = \dot{\varepsilon}_0(E_{21} + E_{22})\frac{\pi d_0^2}{4} \tag{19}$$

Young modulus of the composite may be also expressed by the initial angle of plotted curve - "contact load vs. depth of indentation track" as it follows:

$$\frac{dP}{d\alpha} = \frac{1}{V}\frac{dP}{dt} = \frac{3k}{V}\frac{V}{D}(E_{21} + E_{22})\frac{\pi d_0^2}{4} = \frac{3\pi k}{4}(E_{21} + E_{22})\frac{d_0^2}{D} \qquad (20)$$

Initial angle depends on Young modulus of an elastic component material. It was found that for viscous-elastic components of the composite material the initial angle may be calculated by and be ranged:

$$tg\gamma = \frac{dP}{d\alpha} = 0.8 \quad to \quad 1.3 \qquad (21)$$

Here $tg\gamma$ depends on thickness of the elastic components of the composite material composite material. In present work it was accepted that $tg\gamma=0.8$. Therefore, Young modulus E_{21} of the model (Fig. 1b) is defined by:

$$E_{21} = \frac{3P}{4\sqrt{R}\,\alpha^{3/2}} \qquad (22)$$

Based on experiments on strain relaxation of the composite material it was calculated that $E_{21} = 80$ MPa, $\eta_2 = 8$ MPa·s, and then it was found that $E_{22}=2$ MPa at $\tau = \eta_2/E_{22}= 4$ s.

EXPERIMENTAL TECHNIQUE

The microstructure of the alumina-based composite materials was investigated by X-ray diffraction (XRD), and scanning electron microscopy (SEM) techniques. The chemical composition and structure of the phases and grain boundaries were analyzed by analytical TEM and high-resolution TEM. Conventional and analytical TEM were performed on a 200 kV microscope (model 2000, Japan) equipped with an EDS (Model 6506, England) and a parallel electron energy-loss spectrometer (peels) detector. High-resolution TEM was conducted on a 300 kV microscope (Model 3010, Japan) with a point resolution of less than 0.16 nm. Microhardness was measured with Vickers indentation at load on the indenter of 0.5 N for 30 seconds. Porosity of the oxide hard material was measured by the linear method (method of secant line). The system of texture image analysis "Leitz-TAS" (Germany) was used to study porosity of the materials. Some mechanical parameters of the oxide aluminum composite material are listed in the table 1 below.

To confirm the proposed rheological models Hertzian spherical indentations were done by the indentation techniques at five points. The size of the samples was 25×10×5 mm. They were coated with the following composite systems: 1. "Al$_2$O$_3$ layer – aluminum alloy (Al$_2$O$_3$-Al)"; 2. "CrC – steel base" (CrC-steel); 3. "CrC – Al$_2$O$_3$ layer – aluminum alloy" (CrC-Al$_2$O$_3$-Al). 4. "Steel base – damping viscous-elastic polymer – aluminum alloy – Al$_2$O$_3$ layer".

The indentation was carried out by the spherical steel balls with the diameters of 3.978 mm and 7.978 mm. Five sets of load-unload data were obtained at each point, with the maximum load being increased from $P_1=1$ N to $P_2=2000$ N. For each indentation, unloading was continued up to 5% of the maximum load. After each indentation, the contact radius (a) was measured from the residual contact trace on the top layer. Then, the plot of indentation stress ($\sigma(t)$) versus indentation track depth (α, mm) was obtained and plotted in Fig.5. According to Hertz theory these two parameters should show a linear relationship within the zone.

RESULTS

The Auger spectra obtained at different values of the sputtering time indicated a non-uniform distribution of the substrate components when varying the layer thickness. The Al and O particles in the layer cannot be related to discontinuity of the layer, but the fact of diffusion between the substrate components and the CrC layer during the deposition.

A transition from elastic to fully plastic deformation was observed in the load range investigated. A first order approximate value of the radius of the contact circle at each maximum load was calculated, with the plastic depth determined by differentiating the first part of the unloading curve. A plot of (P) vs. (α) was produced for all five-indentation sets. The "pile up" correction parameter (actual contact radius) was changed in contrast to loading rate, thus indicating different types of deformation. Indentation stress-deformation curves for the Al-Al$_2$O$_3$-CrC composite system and the composite system with polymer intermediate layers are plotted in Fig. 4 that shows indentation depth vs. load curve. It was plotted as a mechanical response of the systems under Hertzian indentation at constant rate with sintered aluminum oxide ball of 4 mm diameter. Lines 1 and 2 show measured experimental data for the Al-Al$_2$O$_3$-CrC composite system and the system with viscous elastic polymer layer respectively. They were compared with experimental data (single line in Fig. 4).

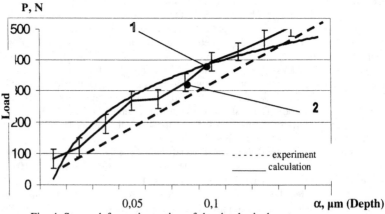

Fig. 4. Stress-deformation rating of the rheological systems

Experimental results were compared with experimental data plotted using approximated single line in Fig. 4 calculated using Eqs. (15-17). The measured values plotted in lines 1, 2 in Fig. 4 are convex down due to effects not taken into consideration for the experiments such as roughness, structural defects etc.; however, the deviation of the results is unexpectedly little.

Viscous elastic polymer structure strongly influences on heterogeneous structure and mechanical properties of the composite system. However, it should be noted that loading conditions, geometry of contacted surfaces and mechanical properties of materials may be included in rheological equations to be used in particular studies or experimental tests. The equations may take difficult mathematical forms and their handmade and quick application in practice would be hard.

At unloading (not shown in Fig. 4) the system has shown the retardation of deformations (elastic return) because of damping viscous-elastic material. It could be shown in the figure as a downfall curve. The plotted relations of experimental and calculated data in Eqs. (15-17) showed good agreement of the developed rheological model and mechanical behavior of the composite systems. Deviation from this line indicates the onset of irreversible deformation. The above stated conditions may be used for investigating mechanical and rheological properties of the "aluminum oxide -aluminum- viscous-elastic material (polymer) – steel" composite systems.

Fig. 5 shows the indentation track obtained on the surface of the CrC-Al_2O_3-Al composite system. The diffuse damage, the traditional cone and ring cracks (Fig. 5b) have been observed in both systems, but the fracture features have affected on the system. The stresses are distributed through top CrC and/or aluminum oxide layers. It was found that stresses concentrate in the zones of pores, internal voids and micro-defects. As a result, it might initiate cracks and fracture of the systems.

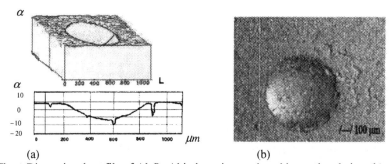

(a)   (b)

Fig. 5. Three Dimensional profile of Al_2O_3-Al indentation track and its sectional view (b), where L is length of profile and α is depth of track, μm, (b) conical and ring cracks in CrC-Al_2O_3-Al.

Fig. 5b shows the surface of the CrC layer. The system has high adhesion to the Al_2O_3 substrate since no exfoliation is visible in the indentation track and the fracture of the coated specimens. The layer is characterized by a fine-grained globule-like structure. Results of the X-ray analysis reveal that the CrC top layer consists generally of chromium carbides (Cr_3C_2 and Cr_7C_3).

The Al_2O_3 layer is generally characterized by numerous micro-failures and micro cracking in the structure. On the other hand, the boundary line of the indentation track on the surface (Fig. 5b) looks uniform and smooth. Of interest is the fact that the micro-hardness near the interface between the aluminum oxide and the CrC layer, which was found to be 20.0 GPa, is higher than that of the uncoated one (about 17 GPa).

The experiments revealed that the $Al-Al_2O_3$-CrC composite system eliminate cracking. CrC nanoparticles hardened structure of the Al_2O_3 layer by filling its pores and healing defects of Al_2O_3 structure. As a result the fracture and crack propagation of the $Al-Al_2O_3$-CrC composite system is suppressed, in contrast to that of the $Al-Al_2O_3$ system. The diameter of the fracture zone and the crack density of the $Al-Al_2O_3$-CrC composite system are significantly smaller then that of the $Al-Al_2O_3$.

By loading the specimen it is found that tensile stress results in localized deformation that is generally elastic-plastic in nature. The deformation is due to the radial spreading of the contact zone under the indenter. This deformation also initiated the local cracks (Fig. 6a). It's assumed that the observed effect of material flows may result in both distribution of stresses and residual deformations of the systems. Fig. 6 shows experimental function of the applied load to the indenter and the stresses of the $Al-Al_2O_3$-CrC system.

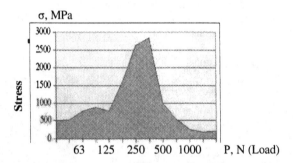

Fig. 6. Applied load-stresses mode of the systems

Shear stresses act in the top ceramic layers at the depth of about 200 μm that is about a half of the thickness of the $CrC-Al_2O_3$ layers. The stresses result in crack initiation and their propagation in the layers. The form and the depth of maximum stress distribution revealed that overall load rating of the system under localized indentation does not depend strongly on hardness of the base substrate. However, it's expected to be important for cycle loading.

In the present research, the ultimate contact stress for the $Al-Al_2O_3$-CrC composite system was about 3 GPa (Fig. 6) that was the highest for the systems under investigations. The ultimate contact stress for the composite system with polymer layers was about 1.5 GPa that was lower for the systems. In fact, polymer layers significantly decrease strength and stiffness of the systems.

CONCLUSIONS

Spherical nano indentation technique is effective technique for measurement contact stress, deformation and effective elastic modulus as a function of penetration depth. The difference in contact pressure and elastic modulus vs. the indenter depth could be adjusted by the modulus vs. the ratio of the contact radius to the film thickness. The errors in the data could be also associated with radial cracks within the contact area. Evidence of residual tensile stresses within the film manifested itself in the form of radial cracks from pores.

A number of mechanisms by which aluminum oxide-based materials may crack or delaminate have been reviewed in this paper. The failure of aluminum oxide-based material is controlled by a number of parameters including the mechanical properties of the material (porosity, defects, voids, and grain structure), interface and substrate, the initial distribution of flaws within the material, and the magnitude and sign of the residual stress in the material. The presence of a tensile stress can cause delamination along the interface, or cracking in the material or substrate. Failure induced by a compressive stress involves a complex interaction between buckling of the material and delaminating along the interface. By healing pores, structural defects and internal voids of aluminum oxide material its fracture resistance and fatigue life might be enhanced in a few times.

ACKNOWLEDGEMENTS

The research works are being supported by the WELCH foundation scholarship, the American Vacuum Society and Dr. F.R. Sheppard - administrator. I would like to acknowledge various assistances of Dr. Sc., Professor Zinovii P. Shulman and Dr. A.A. Mahanek (Institute of Heat and Mass-transfer, National Academy of Sciences of Belarus).

REFERENCES

[1]B.R. Lawn, N.P. Padture, H. Cai, and F. Guiberteau, "Making Ceramics Ductile," *Science*, **263**, 1114-1116 (1994).

[2]B.R. Lawn, "Fracture of Brittle Solids," Cambridge University Press. Cambridge, U.K., 1993.

[3]J.R. Tillett, "Fracture of Glass by Spherical Indenters," *Proc. Ph's. Soc. London*, **8**, 869, 981-992 (1956).

[4]M. Kireitseu, V. Basenuk, and M. Belotserkovskiy, "Composite Bearing Based on Metal-Polymer - Soft Metal-Ceramics Composition," pp. 320-334, Proceedings book of the 104th Annual Meeting & Exposition of the American Ceramic Society, ed. by Dr. Greg Geiger. St. Louis, Missouri, April 29, 2002.

[5]B.R. Lawn, "Hertzian Fracture in Single Crystals with the Diamond Structure," *J. Appl. Phys.*, **39**, 230-245 (1968).

[6]M.T. Laugier, "Hertzian Fracture of Sintered Aluminum oxide," *J. Mater. Sci.*, **19**, 320-325 (1984).

[7]L. An, H.C. Ha, and H.M. Chan, "High-Strength of Aluminum oxide/Calcium Hex aluminum oxides Materials," *J. Am. Ceram. Soc.*, **81** [12] 180-187 (1998).

[8]H. Cai, M. Kalceff, and B.R. Lawn, "Deformation and Fracture of Mica-Containing Glass Ceramics in Hertzian Contacts," *Mater. Res.*, **9** [3] 120-137 (1994).

[9]S.F. Guiberteau, N.P. Padlure and B.R. Lawn, "Effect of Grain Size on Hertzian Contact in Aluminum oxide," *J. Am. Ceram. Soc.*, **77** [7] 450-455 (1994).

[10]L. An, "Fracture and Contact Damage Behavior of Some Aluminum oxide-Based Ceramic," Ph.D. Dissertation. Lehigh Univ. Bethlehem, PA, 1996.

[11]N.P. Padture and B. R. Lawn, "Contact Fatigue of a Silicon Carbide with a Heterogeneous Grain Structure," *J. Am. Ceram. Soc.*, **78** [6] 180-192 (1995).

[12]M.V. Kireitseu, Ion Nemerenco, L.V. Yerakhavets and K.Beltsova. "Nanoindentation of Alumina-Chrome Carbide and Alumina-Ultra Dispersed Diamonds Composite Coatings," pp. 269-277, In: Modeling the Performance of Engineering Structural Materials III, ed. by D.R. Lesuer, T.S. Srivatsan, and E.M. Taleff, TMS 2002 Fall meeting in October 6-10, 2002, Columbus, OH, 2002.

[13]V.A. Rudnitski, A.P. Kren and S.V. Shilko, "Rating the behavior of elestomers by their indentation at constant rates," *Wear*, **22** [5] 502-508 (2001).

[14]Shulman et al. "Rheological behaviour of the composite systems and structures," p. 378, Published by "Nauka", Minsk, 1978.

CHARACTERIZATION OF FeAlN THIN FILMS WITH NANO SIZED PARTICLES

Christophe Daumont[*], Yuandan Liu,
Erik Pavlina, R. E. Miller,
Caspar McConville, Xingwu Wang
Alfred University
Alfred, NY 14802

Robert W. Gray, Jeffrey L. Helfer
BTI
Rochester, NY 14586

Kevin Mooney
SUNY
Buffalo, NY 14260

Peter Lubitz
Naval Research Laboratory
Washington, DC 20375

ABSTRACT

Magnetic FeAlN thin films with nano-sized particles were fabricated via a sputtering process. Chemical compositions of the films were related to the target materials. However, the composition of each film coating was slightly different from that of the corresponding target materials. Moreover, amorphous growth was observed in some of the films. Crystallite sizes were estimated to be less than 10 nm. Magnetic properties of the films are depended on the chemical composition of the film. The saturation magnetization and the remnant magnetization varied in large ranges when the Fe/Al weight ratio varied from 0.1 to 0.9.

INTRODUCTION

Magnetic materials are widely used in electrical, electronic devices and instrumentation. Since the 1990's, progress has been made in magnetic nano-structure fabrication and in the study of their magnetic properties.[1-2] In our previous work, nano-magnetic FeAl, FeAlO and FeAlN films with high Fe/Al ratios were fabricated via a sputtering technique.

In previous work, it was observed that the film composition is different from that of the targets in that for a high iron concentration in the target, a deficiency in iron was found in the resultant film. It was also observed that magnetic properties are related to chemical composition, film thickness and fabrication conditions. In contrast to bulk materials, a thin film material may exhibit different magnetic properties due to the constraint provided by the substrate.[3,4,5] The sputtering fabrication technique is utilized in this work to fabricate nano-magnetic FeAlN thin films as well as AlN films.[3,5,6] This work investigates the relationship between the composition of the target and that of the film, especially with low Fe to Al ratios. Furthermore, this work investigates the relationship between the composition of the thin film and its magnetic properties.

EXPERIMENTS

Fabrication

Nano-magnetic materials were fabricated by a PVD magnetron sputtering process. A Kurt J. Lesker Super System III deposition system outfitted with Lesker Torus magnetrons was utilized for the process (Figure 1).[7]

---

[*]On leave from E.N.S.C.I (French Ceramic Institute).

The vacuum chamber of the system is cylindrical, with a diameter of approximately one meter and a height of approximately 0.6 m. The base pressure is 1-2 μTorr. In these experiments, the targets used are disks with a diameter of approximately 0.07-0.1 m. The sputtering gas is argon with a flow rate of 15-35 sccm. To fabricate the films, a DC power source is utilized at a power level of 500-2,000 W[6] with a pulsed system added in series to provide pulsed DC**. The magnetron polarity switches from negative to positive at a frequency of 100 KHz, while the pulse width for the positive or negative duration can be adjusted to yield suitable sputtering results.

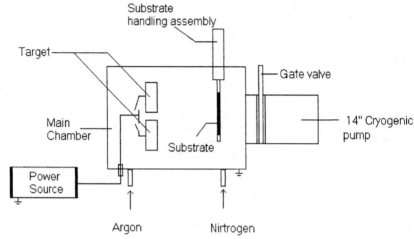

Fig. 1 Sputtering system

The supplier (W: Williams Puretek; L: Lesker Co.) and the weight ratios between Fe and Al for the targets utilized are respectively: 5/95(W), 10/90(L), 20/80(L), 82.5/17.5(L), 90/10 and 95/5(L). Besides argon flowing at a rate of 15-25 sccm, nitrogen is supplied as a reactive gas with a flow rate of 15-30 sccm.[3,5,6]

During fabrication, the pressure is maintained at 2-4 mTorr. This pressure range is found to be suitable for nano-magnetic material fabrication. The substrates are silicon wafers and metal wires. The flat silicon wafers are bare, without a thermally grown silicon dioxide layer, and have a diameter of 0.1-0.15 m. The distance between the substrate and the target is 0.05-0.26 m. During the deposition, the wafer is fixed onto a substrate holder, without rotational motion. A typical wire is a copper wire with a diameter of approximately 0.5 mm. To deposit a film on a wire, the wire is rotated at a rotational speed of 0.01-0.1 rps, and is moved slowly back and forth along its symmetrical axis with a maximum speed of 0.01 m.s^{-1}. To achieve a film deposition rate on the flat wafer of 0.5 nm.s^{-1}, the power required for the AlN films and the FeAlN films is 500 W. A typical film thickness is between 100 nm and 1 μm, and a typical deposition time is between 200 and 2,000 s. In some film structure designs, two layers are fabricated.

**Advanced Energy Sparc-le V, or ENI pulsed power supply.

One is a FeAlN layer with a thickness ranging from 800 to 1,200 nm, and the other is an AlN layer with a thickness ranging from 400 to 700 nm. FeAlN Films on Si wafers have a nominal thickness of 500 nm.

The thickness of the samples is measured by scanning electron microscope (Philips 515 SEM) coupled with x-ray energy dispersive spectroscopy (Evex EDS analytical Si (Li)). To estimate Fe:Al ratios, K-lines obtained in EDS are utilized. The phase formation studies are performed using x-ray diffraction (Philips XRG 3100 XRD system) and transmission electron microscope (JEOL 2000-FX TEM). The magnetic properties of films are determined by superconductive quantum interference device (Quantum Design RF SQUID) magnetometer. A typical measurement procedure to obtain a hysteresis loop for a sample is as follows. The sample is maintained at 300K, and initial magnetization is performed by applying the magnetic field to 2 Tesla. Thereafter, the field is changed from 2 Tesla to −2 Tesla and then back to 2 Tesla to complete a hysteresis loop. (The magnetization of the sample is plotted as function of the applied field).

Material Characterization

SEM-EDS: Figures 2 and 3 show SEM photomicrographs of the cross-sectional areas of two films coated from 10wt%Fe/90wt%Al and 20wt%Fe/80wt%Al targets on copper wire substrates. For each photomicrograph, from left to right, there are three sections: the epoxy, the double layer film (AlN/FeAlN) and the substrate. The lines in the substrate are due to sample polishing prior to observation. From each SEM photomicrograph, the total thickness of the two-layer (FeAlN+AlN) film is between 1 and 2 μm. However, the separation between the two layers is not revealed probably due to interdiffusion between FeAlN and AlN layers during deposition.

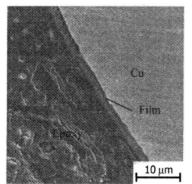 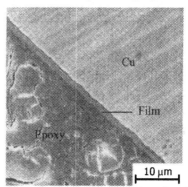

Fig. 2. Thin film fabricated with 20wt%Fe/80wt%Al target on a copper wire substrate.

Fig.3. Thin film fabricated with 10wt%Fe/90wt%Al target on a copper wire substrate.

Figure 4 (a)-(c) are EDS spectra based on film top surface measurement. In Figure 4 (a), EDS spectrum for a copper wire is illustrated. In Figure 4 (b), EDS spectrum for a two layer (FeAlN/AlN) film on a copper wire substrate is illustrated, which is fabricated with 10wt%Fe/90wt%Al target. An Al peak is dominant, along with a copper peak and an indicative peak for Fe. The C and O peaks in Figure 4 (b) are due to sample preparation and handling.

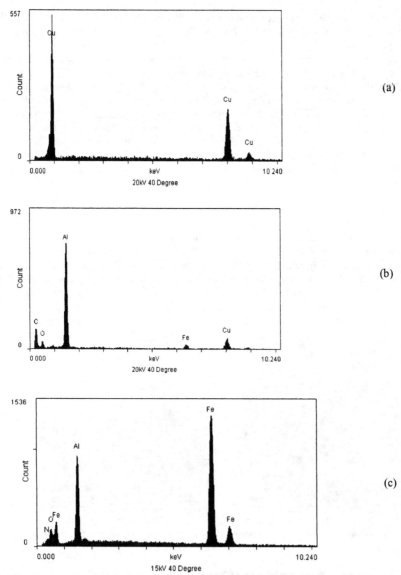

Fig. 4 (a) EDS spectrum of copper wire substrate. (b) EDS spectrum of the FeAlN/AlN film deposition with 10wt%Fe/90wt%Al target on a copper wire substrate. (c) EDS spectrum for a 500 nm FeAlN film fabricated with 82.5wt%Fe/17.5wt%Al target on a Si wafer substrate.

Table I. Relative ratio between Fe and Al determined via K lines

Target composition (wt%Fe)	(wt%Al)	wt%Fe in film	wt%Al in film
82.5	17.5	77	23
20	80	15	85
10	90	12	88
5	95	9	91

XRD/TEM: Figure 5 shows the XRD pattern of a FeAlN film having a thickness of 500 nm on a Si wafer (400) fabricated with 82.5wt%Fe/17.5wt%Al target. There are two broad peaks around 28 and 54 degrees. The average crystallite size is estimated to range from 1 to 2 nm, via a computer program called SHADOW.[8] TEM micrographs of two thin film samples deposited on copper grids (with 82.5wt%Fe/17.5wt%Al and 10wt%Fe/90wt%Al targets) reveal amorphous structures. TEM micrograph of another thin film sample (fabricated with 95wt%Fe/5wt%Al target) reveals nano-crystallite structures, with the crystallite sizes of 2-10 nanometers.

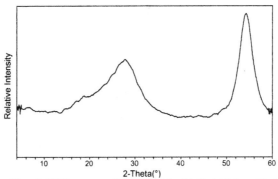

Fig. 5 XRD pattern from a 82.5wt% Fe/ 17.5 wt% Al target

Magnetic Properties

In Table II, sample identifications for 3 samples are tabulated. The films were deposited on copper wires. At 300 K, magnetic hysteresis loops for these samples are plotted in Figures 6-8, via SQUID magnetization measurements. In Figure 6, the hysteresis loop for sample 1 is plotted (The copper substrate contribution to magnetization has been subtracted, and the contribution due to AlN is assumed to be negligible.) This sample shows very weak saturation magnetization, approximately 3 emu·cm^{-3}. Moreover, M_r is approximately 0.05 emu·cm^{-3} and H_c is less than 50 Oe.

In Figure 7, the hysteresis loop for sample 2 is plotted. This plot shows distinctively the properties of superparamagnetism. Under an applied field of 20 000 Oe, the magnetization approximately equals 46 emu·cm^{-3}. Moreover, M_r is approximately 1.4 emu·cm^{-3} and H_c is less than 10 Oe.

Figure 8 indicates a very soft ferromagnetic behavior for sample 3. This sample has a saturation magnetization equal 35 emu·cm^{-3}. Moreover, M_r is approximately 17 emu·cm^{-3} and H_c is approximately 25 Oe.

Table II. FeAlN/AlN films analyzed.

Sample number	Target composition wt%Fe/wt%Al	Nominal thickness (nm)	Magnetization at 2T (emu·cm^{-3})	Remnant magnetization (emu·cm^{-3})
1	Fe$_{10}$Al$_{90}$	800/700	3	0.05
2	Fe$_{82.5}$Al$_{17.5}$	800/400	46	1.4
3	Fe$_{90}$Al$_{10}$	1200/400	35	17

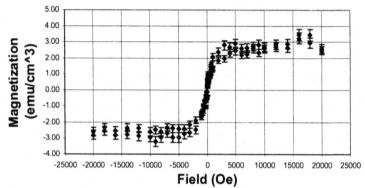

Fig. 6. Hysteresis loop for Sample 1 w/ substrate subtraction

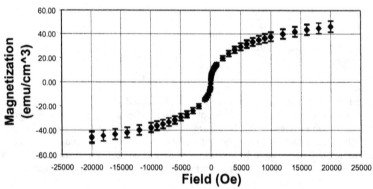

Fig. 7. Hysteresis loop for Sample 2 w/ substrate subtraction

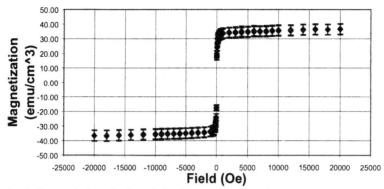

Fig. 8. Hysteresis loop for Sample 3 w/ substrate subtraction

DISCUSSION AND CONCLUSION

Nano-magnetic FeAlN thin films were fabricated via a sputtering process. As revealed by EDS, chemical composition of each film was related to the target materials. However, with experimental error, the composition of the film coating was somewhat different from that of the target materials. Moreover, amorphous growth was observed in some of the films, as shown in XRD and TEM. For two layer (FeAlN/AlN) films, the separation between the FeAlN and the AlN was not revealed probably due to interdiffusion between FeAlN and AlN layers during deposition. Currently, more detailed studies are being carried out to identify the diffusion layer growth mechanism. A low ratio of Fe/Al in the target is found to give an enrichment of iron in the resultant film. Crystallite sizes were estimated to be less than 10 nm. As measured by SQUID magnetization, the magnetic properties of the films depended on the chemical composition of the film. When Fe/Al weight ratio varied from 0.1 to 0.9, the saturation magnetization varies one order of magnitude; and the remnant magnetization varied more than two orders of magnitude. However, the coercive force varied no more than one order of magnitude.

ACKNOWLEDGEMENTS

Work partially supported by Biophan Technologies Inc., NYSTAR-CACT, and NSF-CGR. Help provided by Christopher Norton and Zain Horning.

REFERENCES

[1]A.Carl and E.F. Wassermann, "Magnetic Nanostructures for Future Magnetic Data Storage: fabrication and Quantitative Characterization by Magnetic Force Microscopy," pp.59-92 in *Magnetic Nanostructures*, Edited by S. Nalwa. American Scientific publishers, Stevenson Ranch, California, 2002.

[2]M. Solzi, M. Ghidini, and G. Asti, "Macroscopic magnetic properties of Nanostructured and Nanocomposite Systems," pp. 124-201 in *Magnetic Nanostructures*, Edited by S. Nalwa. American Scientific publishers, Stevenson Ranch, California, 2002.

[3]X.Wang, R. E. Miller, P. Lubitz, F. J. Rachford, J. H. Linn, "Nano-Magnetic FeAl and FeAlN Thin Films via Sputtering," *Ceramic Eng. & Sci. Proc.*, 24 [3] 629-636 (2003).

[4]G.G. Bush, "The Complex Permeability of a High Purity Yttrium Iron Garnet (YIG) Sputtered Thin Film," *J. Appl. Phys.*, 73 [10] 6310-6311 (1993).

[5]X.Wang, R. E. Miller, Y. Liu, R. W. Gray, J. L. Helfer, R. W. Nowak, K. P. Mooney, "Nano-Magnetic Coatings on Metallic Wires," in *Proceedings of 52nd IWCS*, pages 647-653.

[6]X.Wang, R. E. Miller, J. Linn, C. Washburn, "Technique Devised for Sputtering AlN Thin Films," p 35, the GlassResearcher Vol. 12, No 1 and No 2, Fall 2002 and Spring 2003.

[7]The Kurt J. Lesker Company, Clairton, PA. The magnetron is Torus 4.

[8]S.A Howard, "SHADOW: A system for X-ray powder diffraction pattern analysis; Annotated program listings and tutorial", University of Missouri-Rolla, 1990.

TEM STUDY OF NANOSTRUCTURED MAGNESIUM ALUMINATE SPINEL PHASE FORMATION

Jafar F. Al-Sharab,
Ceramics and Materials Engineering
Rutgers University
Piscataway, NJ 08904, USA

Amit Singhal and Ganesh Skandan
NEI Corporation
Suite 102/103, 201 Circle Drive
Piscataway, NJ 08854-3723, USA

James Bentley
Metals and Ceramics Division
Oak Ridge National Laboratory
Oak Ridge, TN 37831-6064, USA

Frederic Cosandey
Ceramic and Materials Engineering
Rutgers University
Piscataway, NJ 08904, USA

ABSTRACT

The direct conversion of nanostructured $MgAl_2O_4$ spinel from γ-Al_2O_3 has been investigated using three different initial nanostructured powders with differing particle size and agglomerate size. The effect of post-annealing temperature (300, 500 and 800°C) on $MgAl_2O_4$ phase formation was investigated using Transmission Electron Microscopy (TEM) and Selected Area Electron Diffraction (SAED). Relative diffraction intensities as well as lattice parameter measurements revealed that $MgAl_2O_4$ spinel structure starts forming at temperatures as low as 300°C. Further annealing to 800°C and 850°C leads to pure spinel phase with limited particle growth. Data suggests that pure $MgAl_2O_4$ phase formation is favored by small initial γ-Al_2O_3 particle size as well as small primary agglomerate size.

INTROUDUCTION

Magnesium aluminate spinel ($MgAl_2O_4$) as well as other spinel materials such as chromites ($MgCr_2O_4$) and magnetite (Fe_3O_4) have receiving a lot of attention due to their unique combination of physical, chemical, optical, electrical and magnetic properties [1, 2]. With a high melting temperature of 2135°C [1] and resistance to chemical attack [1, 3], $MgAl_2O_4$ has been used extensively as an excellent refractory material in harsh environments [3]. Spinel is also used in optical devices [1, 2] such as passive Q-switch of lasers and for electronic applications such as humidity sensors [2, 4]. In order to achieve ultimate optical properties with excellent mechanical integrity, lots of studies have been devoted in recent years towards the synthesis of nanostructured spinel powders with very narrow size distribution[4].

At present, most techniques used to produce $MgAl_2O_4$ spinel, whether they are solid-state reactions or chemical synthesis, are conducted at high temperatures (>800°C) [2-4]. However, it is known that high temperature processing is detrimental to microstructure properties due to non-uniform grain growth. In recent years, low temperature chemical synthesize of nano crystalline powders of $MgAl_2O_4$ spinel has been explored. Recently, our group has prepared nanostructured $MgAl_2O_4$ spinel by a direct conversion process form γ-Al_2O_3 at low temperatures (<300°C). Details of this synthesis process will be published elsewhere [5].

This paper presents results and characterization by TEM of $MgAl_2O_4$ synthesized at low temperatures. Three initial types of γ-Al_2O_3 powders were used with different primary particle size and with varying degree of agglomeration. Dark field TEM as well as selected area electron diffraction (SAED) were used in this investigation.

EXPERIMENTAL

Three commercially available γ-Al_2O_3 nano powders with different particle characteristics were utilized to synthesize nano-$MgAl_2O_4$ powders. Information on primary particle size, agglomerate size and surface area for these γ-Al_2O_3 nano powders are given in Table I. The surface area and agglomerate size were estimated according to the Brunaure-Emmett-Teller (BET) model and Photon Correlation Spectroscopy (PCS) respectively. The particle size was measured from dark field TEM images. Details of the synthesis processes will appear elsewhere [5], in brief equal molar fraction of magnesium nitrate and γ-Al_2O_3 nano particles were mixed and treated for prolonged period (~16 hrs) at 270°C. Then the powders were heat treated in air at temperatures of 800°C for sample A and 850°C for samples B and C to fully decompose the excess nitrate and facilitate the formation of the spinel phase. For powder A, the effect of progressive heating temperature of 300, 500 and 800°C was also investigated. TEM samples were prepared by dispersing the powder in trichlorotrifluoroethane, followed by placing a drop of the suspension on holey carbon film supported on copper grids. For TEM and SAED analysis, we used a TOPCON 002B microscope equipped with Gamma-PGT x-ray detector. Simulations of SAED patterns were performed using JEMS software [6]. Grain size analyses were obtained by measuring the crystallite size from dark field images. More than 120 particles were measured in each sample.

Table I. Surface area, initial agglomerate and particle sizes of various starting γ-Al_2O_3 powders

Sample ID	Surface Area (BET) (m^2/g)	Agglomerate size (PCS) (nm)	Particle size (TEM) (nm)
A	120	120	14
B	120	400	11
C	30-60	150-160	40

RESULTS and DISCUSSION
Effects of processing on particle size analysis

In order to follow the effect of processing temperature on the final particle size, TEM dark field images were taken from all three samples A, B and C. Typical dark field TEM images of initial γ-Al_2O_3 (Sample A), $MgAl_2O_4$ processed and annealed at 800°C are given in figures 1a and 1b respectively. The grain size distributions measured from these two previous conditions are given in figures 2a and 2b with average size for various processing conditions summarized in Table II. Grain size analysis shows only a slight increase in particle size from 14 nm to 17 nm after processing and annealing at 800°C.

The TEM dark field images of samples B and C processed and annealed at 850°C are shown in Figures 3a and 3b respectively. As to be expected from the starting γ-Al_2O_3 powders, these dark field images show significant difference in crystalline size. Sample B has a much small crystalline size, while sample C has a larger crystals. Grain size distributions of these two samples are also displayed in figures 4a (Sample B) and 4b (Sample C) were it is observed that the average particle size of Sample B is smaller (11 nm) with narrow size distribution (5-25 nm). On the other hand, sample C shows a larger average particle size (37 nm) with a wide distribution (5-62 nm). As for sample A, the material does not experience significant particle growth after processing at 850°C

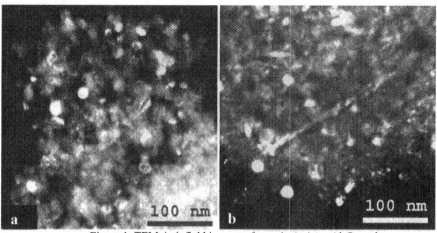

Figure 1. TEM dark field images of sample A, (a) γ-Al₂O₃ and
(b) MgAl₂O₄ after processing and annealing at 800°C.

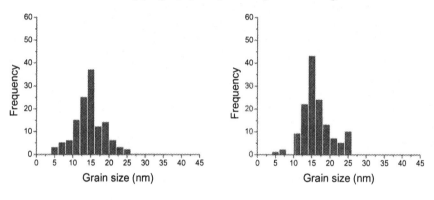

(a)   (b)

Figure 2 Grain size distribution of sample A, (a) as synthesized γ-Al₂O₃ and
(b) MgAl₂O₄ after processing and annealing at 800°C

Table II. Grain size measurements of as synthesized γ-Al₂O₃ (Sample A)
and MgAl₂O₄ after annealing at various temperatures

Material	Avg. crystalline size (nm)
γ-Al₂O₃	14.0 ±2
300°C	14.9 ±2
500°C	17.0 ±2
800°C	16.8 ±2

Figure 3. TEM dark field images of MgAl$_2$O$_4$ obtained from sample B (a) and sample C (b) after processing and annealing at 850°C.

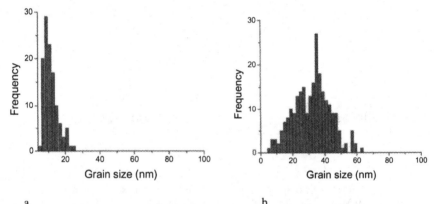

a   b

Figure 4. Grain size distributions of MgAl$_2$O$_4$ obtained from sample B (a) and sample C (b) after processing and annealing at 850°C.

Structure and diffraction simulations of γ-Al$_2$O$_3$ and MgAl$_2$O$_4$ spinel

Aluminum Oxide (Al$_2$O$_3$) exists in many polymorphous phases depending on the distribution of cations and anions in the structure [7]. For nano-structured powders, Al$_2$O$_3$ usually exist in the form of cubic γ-Al$_2$O$_3$ [7, 8] with space group, Fd-3m, and atomic positions and site occupancy summarized in table III. The MgAl$_2$O$_4$ spinel has a similar structure with atomic positions and occupancy for Mg, Al and O given in Table III. The O atoms form a FCC sub-lattice with Mg atoms occupying tetrahedral interstitial sites while Al atoms occupying octahedral interstitial sites. Simulation of polycrystalline diffraction patterns for γ-Al$_2$O$_3$ and MgAl$_2$O$_4$ spinel are displayed in Figures 5 and 6 respectively.

A comparison of the two diffraction patterns of MgAl$_2$O$_4$ spinel and γ-Al$_2$O$_3$ shows two major differences. In the MgAl$_2$O$_4$ spinel, the {111} and {333} reflections are more intense with even the {111} absent from γ-Al$_2$O$_3$. In addition, the {331} reflection decreases in intensity upon the transformation from γ-Al$_2$O$_3$ to MgAl$_2$O$_4$. Based on the diffraction simulations shown above, it is clear that diffraction technique is critical to determine the existence and evolution of spinel from γ-Al$_2$O$_3$, primarily from the presence and relative intensity of the {111} reflection.

Table III. Crystal structures of γ-Al$_2$O$_3$ [9] and MgAl$_2$O$_4$ phases [10].

γ-Al$_2$O$_3$ Space group Fd-3m , a$_o$ = 0.7906 nm				MgAl$_2$O$_4$ Space group Fd-3m , a$_o$ = 0.8083 nm			
Atoms	X,Y,Z	Occupancy	No. Atoms	Atoms	X,Y,Z	Occupancy	No. Atoms
O	0.245	1	32	O	0.262	1	32
Al$_1$	0.125	1	8	Mg	0.125	1	8
Al$_2$	0.5	0.833	13.3	Al	0.5	1	16

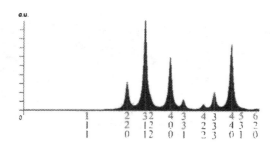

Figure 5 Simulated electron diffraction intensities of γ-Al$_2$O$_3$ with 15 nm grain size.

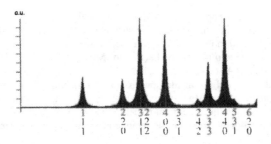

Figure 6 Simulated electron diffraction pattern of MgAl₂O₄ spinel with 15 nm grain size.

Analysis of selected area electron diffraction (SAED) patterns

       Selected area electron diffraction were obtained for the starting γ-Al$_2$O$_3$ powder as well as after processing and annealing in order to follow the formation of the MgAl$_2$O$_4$ spinel phase. SAED pattern of as synthesized γ-Al$_2$O$_3$ sample A, as well as processed and annealed at 800°C are shown in Fig. 7a and 7b respectively. The SAED pattern obtained after processing and annealing at 800°C shows clearly an additional peak at d=0.4662 nm corresponding to the {111} reflection of MgAl$_2$O$_4$ spinel. As expected, such a reflection is missing from the as synthesized γ-Al$_2$O$_3$ powder. The diffraction pattern is also less diffuse with discernable spots that are indicative of the presence of larger particles. This results is consistent with the measured increase in grain size (cf. Table II)

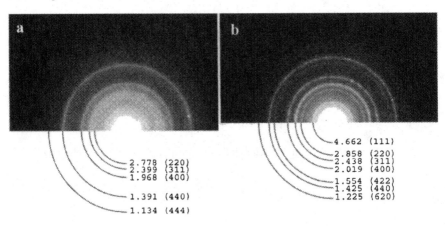

Figure 7. SAED patterns of sample A, as synthesized γ-Al$_2$O$_3$ (a) and processed and annealed at 800°C (b), (d-spacing in Angstrom).

Figures 8a and 8b show the diffraction patterns of samples B and C after processing and annealing at 850°C. Both patterns show the {111} reflection. In addition, Sample B shows diffuse continuous rings, while sample C has a rings made of spots reflecting the larger grain size distribution present in this sample. Furthermore, sample C contains additional reflections that are not part of the spinel phase. These reflections belong to the MgO phase. Other MgO reflections overlap well with $MgAl_2O_4$ spinel reflections. The presence of MgO in this sample indicates that certain amount of Mg had not reacted with γ-Al_2O_3 due to the initial larger particle size and then transformed to MgO upon annealing in air

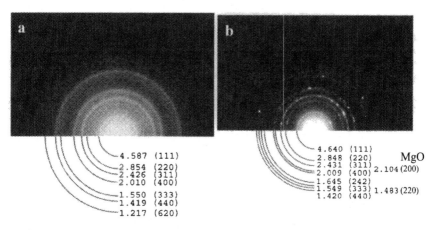

Figure 8. SAED patterns of (a) samples B and (b) sample C after processing and annealing at 850°C.

In order to analyze in more details the spinel phase formation, SAED patterns from the various samples and annealing conditions were analyzed further to extract the relative intensities of the major reflections (111, 220, 311, 400, 333 and 440). For this purpose, the intensity was first measured for each ring using rotational average method [11]. For this analysis, the Process-Diffraction program was used [11]. Then the background intensity was removed with power fit law function of degree 3. Similar processes were conducted for all SAED patterns. Figure 9 compares the extracted intensity profiles of sample A, as synthesized and processed at various temperatures of 300, 500 and 800°C. Intensity profiles show that spinel starts to form already at 300°C from the presence of a small {111} peak and continues to grow until the final temperature of 800°C is reached. The {333} reflection also shows an increase in intensity with respect to {311} reflecting spinel phase formation (see figures 5 and 6). Also, there is a shift in peak positions to the left indicative of an expansion in the lattice parameter. This expansion in lattice parameter was calculated by measuring the lattice parameter from the {220} reflection and the results are presented in figure 10. These results show that there is a gradual increase in lattice parameter upon annealing temperature which parallel the gradual increase in {111} peak intensity. The large dispersion in the data points at 300°C can be related to inhomogeniety between various areas of the sample. However, annealing at 800°C show more homogeneity as lattice parameters measurements indicates. Moreover, EDS and EELS analysis have also shown that sample processed and annealed at 800°C exhibit more homogeneity than sample processed and annealed at 300°C [12] . The lattice parameter measurements show that there is about 8.6 % increase in lattice volume from as synthesized γ-Al₂O₃ to annealed sample at 800°C. This is consistent with the change in lattice parameter between synthesized γ-Al₂O₃ with a lattice parameter of 0.790 nm [7] and MgAl₂O₄ with lattice parameter of 0.8083 nm [3] resulting in 7.1% increase in volume.

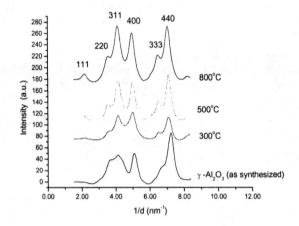

Figure 9. SAED intensity profile comparison of as synthesized γ-Al₂O₃ sample A and MgAl₂O₄ after processing and annealing at 300, 500 and 800°C

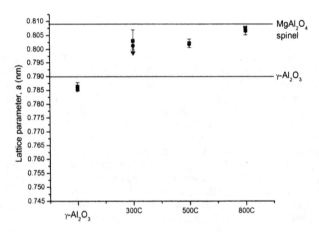

Figure 10. Lattice parameter evolution of γ-Al₂O₃ to spinel as a function of temperature

The SAED intensity profiles of samples B and sample C are shown in Figure 11 revealing the presence of {111} reflection. As presented earlier, this is an indication of the existence of spinel phase. Spinel from sample B has broader peaks with lower {111}/{311} and {333}/{440} relative intensity ratios. However, sample C exhibits well-defined peaks and higher {111}/{311} and {333}/{440} relative intensity ratios. Extra peaks appearing in the spectra match well with {200} and {220} reflections of the MgO material as mentioned earlier. The presence of this phase may be related to the reaction of extra Mg with O to form the MgO phase during annealing the powders in ambient oxygen. X-ray data also show the presence of minute amounts of MgO in samples A and B as revealed from an asymmetry of the base of {440} and {400} reflections [12].

Although both samples B and C were annealed to the same temperature (850°C), sample B may not be fully transformed to spinel as observed from the lower relative diffraction intensity of the {111} reflection (figure 11). This could be explained in part by the fact that in sample B the particles are agglomerated. As the transformation from γ-Al$_2$O$_3$ to MgAl$_2$O$_4$ spinel is controlled by diffusion of Mg into tetrahedral interstitial sites, the transformation into spinel is incomplete within the interior of the aggregate of sample B. Lattice parameter (a$_o$) measurements also confirm this observation, as lattice parameters of samples B and C are 0.7973 and 0.8037 nm respectively.

For sample C the transformation to spinel phase appears complete but possesses a large amount of MgO. As explained earlier, this may be related to the large initial particle size (cf. Table I).

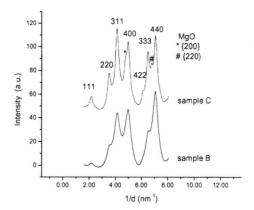

Figure 11. SAED Intensity profile comparison of sample B and sample C
after processing and annealing at 850 °C

CONCLUSIONS

Nanostructured MgAl$_2$O$_4$ spinel was synthesized at low temperatures by direct conversion from γ-Al$_2$O$_3$. Although, increasing annealing temperature from 300 to 800°C is accompanied with full transformation to MgAl$_2$O$_4$ spinel with an associated increase in lattice volume, the initial particle size of as synthesized γ-Al$_2$O$_3$ shows limited growth. The γ-Al$_2$O$_3$ transformation into MgAl$_2$O$_4$ spinel is facilitated by both small initial particle and agglomerate sizes.

ACKNOWLEDGMENTS
The project was supported in part by Small Business Innovation Research sponsored by the US Army. Research at the Oak Ridge National Laboratory SHaRE User Center was sponsored by the Division of Materials Sciences and Engineering, U.S. Department of Energy, under contract DE-AC05-00OR22725 with UT-Battlelle, LLC.

REFERENCES

1. J.-G. Li, T. Ikegami, J.-H. Lee, T. Mori, and Y. Yajima, *A wet-chemical process yielding reactive magnesium aluninate spinel (MgAl$_2$O$_4$) powder.* Ceramics International, **27**. 481-89 (2001)

2. E.H. Walker, J.W. Owens, M. Etenne, and D. Walker, *The novel low temperature synthesis of nanocrystalline MgAl$_2$O$_4$ spinel using "gel" precussors.* Materials Research Bulletin, **37**. 1041-50 (2002)

3. L.R. Ping, A.-M. Azad, and T.W. Dung, *Magnesium aluninate (MgAl$_2$O$_4$) spinel produced via self-heat-sustained (SHS) technique.* Materials Research Bulletin, **36**. 1417-30 (2001)

4. G. Gusmano, P. Nunziante, T. Traversa, and G. Chiozzini, *The mechanism of MgAl$_2$O$_4$ spinel formation from the thermal decompostion of coprecipitated hydroxides.* Journal of the European Ceramic Society, **7**. 31-39 (1991)

5. A. Singhal, G. Skandan, J.F. Al-Sharab, and F. Cosandey, *A new low temperature processes for synthesizing multicompound and doped oxide nanoparticles.* To be published, (2004)

6. P. Stadelmann, *http://cimewww.epfl.ch/people/stadelmann/jemsntv1_3402w2003.htm.*

7. I. Levin and D. Brandon, *Metastable alunina polymorphs: crystal structures and transition sequences.* Journal of American Ceramic Society, **81**[8]. 1995-2012 (1998)

8. L. Fu, D.L. Jonson, J.G. Zheng, and V.P. Dravid, *Microwave plasma synthesis of nanostructured γ-Al$_2$O$_3$ powders.* Journal of American Ceramics Society, **86**[9]. 1635-37 (2003)

9. L. Deyu, B.H. O'Cornnor, G.I.D. Roach, and J.B. Cornell, *structural models of eta and gamma alunias by x-ray retveled refinement.* Acta Crystallographica A, **39** (1983)

10. P. Villars and L.D. Calvert, *Pearson's Handbook of Crystallographic Data for intermetallic phases.* ASM international, **1**. 906 (1991)

11. J. Lábár, *"ProcessDiffraction": A computer program to process electron diffraction patterns from polycrystalline or amorphous samples.* Proceedings of EUREM 12, Brno (L Frank and F Ciampor, eds.), Czechoslovak Society for Electron Microscopy. I379-I80 (2000)

12. J.F. Al-Sharab, A. Singhal, G. Skandan, J. Bentely, and F. Cosandey, *Synthesis and Characterization of nanostructured MgAl$_2$O$_4$ spinel materials.* To be published, (2004)

Nanotubes and Nanorods

GROWTH OF CARBON NANOTUBES BY MICROWAVE PLASMA CHEMICAL VAPOR
DEPOSITION (MPCVD)

L. Guo, V. Shanov, and R. N. Singh
Department of Chemical and Materials Engineering
University of Cincinnati
P.O. Box 210021
Cincinnati, OH 45221-0012

ABSTRACT
    Carbon nanotubes were grown on Si or oxidized Si substrates using cobalt or nickel
catalysts and methane by microwave plasma chemical vapor deposition (MPCVD). Effects of
plasma gases, catalyst thin films and substrates on morphology of carbon nanotubes were
studied. In-situ mass-spectroscopy of the gaseous phase was used to identify the chemical
reactants and products during MPCVD. The nanotubes were characterized by SEM and Raman
spectroscopy.

INTRODUCTION
    Carbon nanotubes (CNT) are graphitic tubes in diameters of nanometers, lengths up to
several microns, and have very high aspect ratios. They exhibit highly anisotropic physical and
chemical properties that are promising for applications in one-dimensional electronic devices
and technologies. Several methods were used to prepare carbon nanotube with different
structures and morphology, including arc discharge, laser vaporization, and pyrolysis. To
realize their potentials, controlled growth of well-aligned nanotubes is essential. Chemical
Vapor Deposition (CVD) is one of the promising methods to control the growth process
through control of the substrate temperature and carbon concentration. Another advantage of
CVD approach is that CNTs can be made continuously and thus if the optimum conditions for
growing pure CNTs could be found, it is a very good way to synthesize large quantities of
CNTs under relatively controlled conditions.
    Recently, considerable progress has been made in fabrication of multiwall nanotubes by
microwave plasma chemical vapor deposition (MPCVD).[1-4] Ammonia has been shown to
stabilize the alignment of carbon nanotubes on nickel,[5] but twist-like defects also exist along
carbon nanotubes. However, ammonia is not necessary for formation of CNTs, the influence of
nitrogen in plasma gas sources on formation and growth mechanism of carbon nanotubes has
also been investigated. First, as the amount of nitrogen increases, the alignment of the CNT
lattice is gradually lost;[6] second, nitrogen has been found either to enhance the formation of
graphite layer on the catalyst surface or to increase the separation of the graphitic layers from
the catalyst;[7] It was also found that the presence of nitrogen in plasma gases would enhance the
carbon diffusion into the catalyst, and help to form bamboo-like structures.[4]
    In this work, the role of nitrogen in plasma was further studied by varying gas
concentrations. In-situ mass-spectroscopy of the gaseous phase was used to identify the
chemical reactants and products during MPCVD, which may provide some information on the
growth mechanism of CNTs. On the other hand, different types of substrates and catalysts were
also used in order to investigate the optimum parameters for CNT growth in nitrogen
containing plasma.

EXPERIMENTAL

An ASTEX microwave plasma deposition system (see in Figure 1) was used to prepare carbon nanotubes (CNTs). Before deposition, nickel (Ni) and cobalt (Co) thin films were deposited on silicon and oxidized silicon by a magnetron sputtering system. Then the coated samples were loaded into the chamber. The chamber was degassed in vacuum (10^{-6} torr) in order to remove the water vapor on the walls. Then the samples were heated in hydrogen (H_2) flow at 700 °C and pressure of 20 torr in order to enhance the adhesion of catalyst with the substrates, and to break the continuous catalyst films into nano-sized islands. Carbon nanotubes were formed in two steps at 700 °C. In the first step, the samples were exposed to hydrogen plasma under a flow rate of 100 sccm, at an operation pressure of 20 torr and a microwave power of 600 W. In the second step, nitrogen (N_2) and methane (CH_4) were introduced, and the concentration of nitrogen was varied from 0% to 80%, and methane concentration was fixed at 20% of the flow rate of 100 sccm. After 30 minutes, the chamber was cooled down in hydrogen. The substrate temperature was monitored by a thermocouple buried under the sample stage and the plasma temperature was measured by an infrared pyrometer.

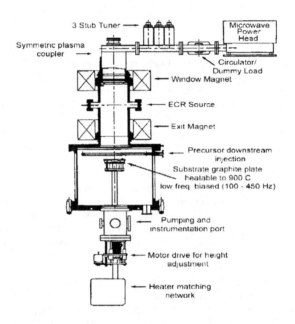

Fig. 1. Schematic diagram of the microwave plasma CVD system.

In-situ mass-spectroscopy (MKS Instruments Orion Compact) of the gaseous phase was used to identify the chemical reactants and products during MPCVD, which provided some information on growth species during the deposition process. Temperature of the plasma ball was monitored by an infrared pyrometer (Raytek Thermalert GP). The morphology of carbon nanotubes was examined by scanning electron microscopy (SEM, Hitachi S4000). Raman spectroscopy was done using Ar$^+$ laser with a wavelength of 514.5 nm (Micro-Raman T64000 Jobin Yvon triple monochromator system).

RESULTS AND DISCUSSION

It is known that catalytic growth of CNTs involves absorption of carbon species from the gas phase and precipitation of carbon from the catalyst solution. There might be more process steps including ionization of the plasma gas sources, and changes in catalyst thin film in plasma. In order to investigate the effects of nitrogen content on these catalytic reactions, CNT growth using different nitrogen concentration (0, 20, 80 vol %) in the plasma gas were investigated.

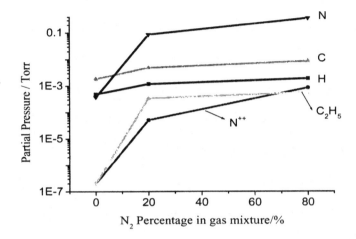

Fig. 2 Active species detected in plasma during deposition of CNTs using quadrupole mass spectroscopy (QMS)

Figure 2 shows changes in reactive species in plasma with different nitrogen content. It is clear that the partial pressure of nitrogen related species (such as N, N^{++}) increases with nitrogen concentration. This increase may give rise to greater ion bombardment effect compared to the hydrogen plasma, since nitrogen atoms and ions are heavier and more energetic. This also increases the plasma ball temperature, which was observed during the deposition. The infrared pyrometer readings showed 780 °C for 0% nitrogen, 830 °C for 20% nitrogen and over 1100 °C for 80% nitrogen. This temperature increase may melt and agglomerate more nanoparticles into bigger sizes, which is shown in Figure 3 (d, e and f). On the other hand, some complex species (such as C$_2$H$_5$) were observed in nitrogen plasma.

The effects of nitrogen content on CNT growth is also shown in Figure 3. SEM images of nickel coated substrates (a, b and c) show that the diameter and length of CNT products increases from 20nm to 100 nm with nitrogen content, indicating that most of them may be multiwall nanotubes. This conclusion is expected to be configured by HRTEM (high resolution

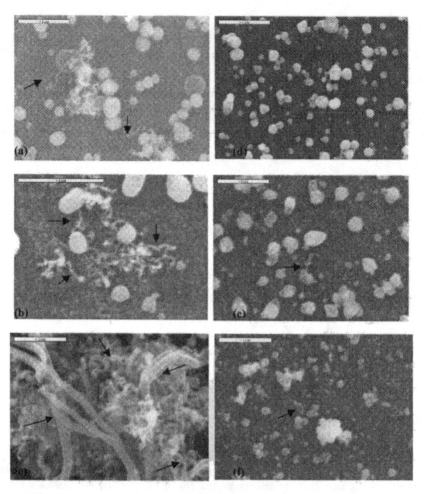

Fig. 3 SEM images of catalyst on oxidized Si substrates at T=700 °C, W=900W, P=20torr and CH$_4$=20% for 30 minutes. (a) Ni, 80% H$_2$; (b) Ni, 60% H$_2$, 20% N$_2$; (c) Ni, 80% N$_2$; (d) Co, 80% H$_2$; (e) Co, 60% H$_2$, 20% N$_2$; (f) Co, 80% N$_2$.

transmission electron microscopy) in the near further. Big clusters of CNTs were found at high nitrogen concentration. One possible reason is the increased temperature of the plasma ball causing melting or agglomeration of the nano-catalyst particles, which would influence the morphology of CNTs growing from them.[2] Another reason may be related to a higher degree of the methane decomposition and creation of carbon related active species in nitrogen plasma (such as C, CH, and CH$_3$) under high nitrogen content. This is shown in Figure 4. The partial pressure of methane decreases with nitrogen concentration, because methane decomposes into species such as C, CH, or CH$_3$. It is not clear which species are dominant to form CNTs. We assume that the more decomposed carbon species, the more of them get dissolved into the catalyst, which enhances the CNT growth.

In Figure 3, it was also noticed that with different catalysts, the results of CNT formation was quite different. Tube-like structures are pointed by arrows in the figure. Carbon nanotubes are more easily formed on nickel coated substrates compared to those on cobalt under the same experiment conditions. The growth mechanism under this condition has not been clarified yet.

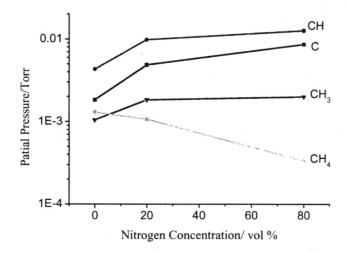

Fig. 4 Active carbon related species detected in the plasma by using QMS

The nature of substrates also has an effect on the CNT production. Figure 5 shows that using oxidized silicon substrates helps to prevent catalysts from reacting with the substrates. On silicon substrate, both nickel and cobalt form silicides (such as NiSix, CoSix, etc.) under high temperature and power, which makes the catalysts to loose their efficiency for formation of CNTs (see Figure 5(b)). The distribution of catalyst particles on silicon substrates is not as uniform as that on oxidized silicon, which is also shown in Figure 5.

Figure 6 shows the Raman spectroscopy. The G-peak (1568 cm^{-1}), D-peak (1348 cm^{-1}) and several peaks in the range of 400-1000cm^{-1} were detected. The strong peak at 1568 cm^{-1} (G-line) can be assigned to the Raman allowed phonon mode E$_{2g}$, which involves out-of-phase interlayer displacements, indicating a graphitic sheet structure. No radial breathing mode (RBM) peak (190.5 cm^{-1}, 215.4 cm^{-1}) appeared in the spectrum, which could suggest that there are no single-wall nanotubes in the CNT products. There is an indication of a peak around 1614 cm^{-1} and no second order peak near 1720 cm^{-1} is observed which suggests multiwall nanotubes.[8] The D-peak at 1348 cm^{-1} is also observed in the spectra. This is attributed to the carbonaceous particles, defects in the curved graphene sheet, nanotube ends, and the finite size of the crystalline domains of the nanotubes. Peaks located in the range of 400-1000 cm^{-1} indicate finite length of CNTs.

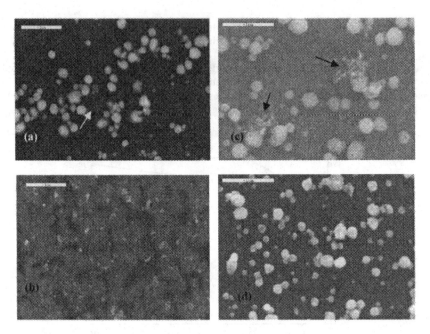

Fig. 5 SEM images of catalyst coating on different substrates at T=700 ºC, W=900W, P=20torr, H₂=80% and CH₄=20% for 30 minutes. (a) Ni coated oxidized Si substrate; (b) Ni coated Si substrate; (c) Co coated oxidized Si substrate; (d) Co coated Si substrate.

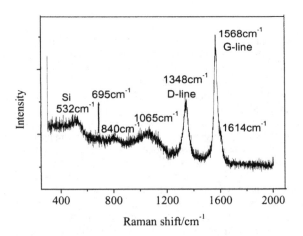

Fig. 6 Raman spectra of carbon nanotubes synthesized with MPCVD at T=700 ºC, W=900W, P=20torr, H₂=80% and CH₄=20% for 30 minutes.

CONCLUSIONS

Carbon nanotubes have been grown using methane/hydrogen/nitrogen mixture using a MPCVD approach. Effective growth of CNTs was found at the temperature of 700 °C, pressure of 20 torr, plasma power of 900 W, methane concentration of 20%, and ratio of nitrogen to hydrogen of 1 to 3. Multiwall CNTs with diameter from 20 to 100 nm and length up to 4 μm were obtained and characterized by Raman spectra and SEM. Nitrogen plasma increased the decomposition of methane, which essentially influenced the morphology of the nanotubes. Oxidized silicon substrates may be more effective for growing carbon nanotube by avoiding formation of the metal silicides in MPCVD environment.

ACKNOWLEDGEMENTS

We thank Dr. Ray Lin and Ms. Yimin Wang for magnetron sputtering of catalysts; Dr Punit Boolchand and Dr. Sergey Mamedov for Raman spectroscopy measurements; Mr. Srinivas Supramanian for SEM and Mr. Rahul Ramamurti, Mr. Ratan S. Kukreja and Mr. Vidhya S. Jayaseelan for helping with experiments. This work is supported by National Science Foundation through the grant No.ECS-0210283. Any opinions, findings and conclusions or recommendations expressed in this material are those of the authors and do not necessarily reflect the views of the National Science Foundation.

REFERENCES

[1] C. Bower, W. Zhu, S. Jin and O. Zhou, "Plasma-induced Alignment of Carbon Nanotubes," *Appl. Phys. Lett.*, **77** [6] 830-32 (2000).

[2] C. Bower, O. Zhou, W. Zhu, D. J. Werder and S. Jin, "Nucleation and Growth of Carbon Nanotubes by Microwave Plasma Chemical Vapor Deposition," *Appl. Phys. Lett.*, **77** [17] 2767-69 (2000).

[3] Y. C. Choi, Y. M. Shin, Y. H. Lee, B. S. Lee, G. S. Park, W. B. Choi, N. S. Lee and J. M. Kim, "Controlling the Diameter, Growth Rate, and Density of Vertically Aligned Carbon Nanotubes Synthesized by Microwave Plasma-enhanced Chemical Vapor Deposition," *Appl. Phys. Lett.*, **76** [17] 2367-69 (2000).

[4] C. H. Lin, H. L. Chang, C. M. Hsu, A. Y. Lo and C. T. Kuo, "The Role of Nitrogen in Carbon Nanotube Formation," *Diamond and Related Materials*, **12**, 1851-57 (2003).

[5] Z. F. Ren, Z. P. Huang, J. W. Xu, J. H. Wang, P. Bush, M. P. Siegal and P. N. Provencio, "Synthesis of Large Arrays of Well-aligned Carbon Nanotubes on Glass," *Science*, **282**, 1105-07 (1998).

[6] R. Kurt and A. Karimi, "Influence of Nitrogen on the Growth Mechanism of Decorated C: N Nanotubes," *Chemphyschem*, **6**, 388-92 (2001).

[7] M. Jung, K. Y. Eun, Y. J. Baik, K. R. Lee, J. K. Shin and S. T. Kim, "Effect of NH_3 Environmental Gas on the Growth of Aligned Carbon Nanotube in Catalytically Pyrolizing C_2H_2," *Thin Solid Films*, **398-399**, 150-55 (2001).

[8] C. J. Lee, D. W. Kim, T. J. Lee, Y. C. Choi, Y. S. Park and W. S. Kim, "Synthesis of Uniformly Distributed Carbon Nanotubes on a Large Area of Si Substrates by Thermal Chemical Vapor Deposition," *Appl. Phys. Lett.*, **75** [12] 1721-23 (1999).

HEAT TREATMENT EFFECT ON THE STRUCTURE OF TiO$_2$-DERIVED NANOTUBES PREPARED BY HYDROTHERMAL METHOD

Yoshikazu Suzuki, Singto Sakulkhaemaruethai, Ryuhei Yoshida and Susumu Yoshikawa
Institute of Advanced Energy, Kyoto University, Uji, Kyoto 611-0011, Japan

ABSTRACT

TiO$_2$-derived nanotubes obtained by the hydrothermal method have been attracted much attention since the innovative work by Kasuga et al. in 1998, because of their fascinating microstructure and promising photoelectrochemical applications. Some recent studies suggest that nanotubes by this method contain a fair amount of water, implying that they should have a layered titanate structure. In this study, heat treatment effect on the structure of TiO$_2$-derived nanotubes has been investigated. High-temperature XRD and TG-DTA analysis revealed the structural change of the TiO$_2$-derived nanotubes. A preliminary *in situ* high-temperature SEM study also demonstrated their microstructural change during heating.

INTRODUCTION

Since the innovative work by Kasuga et al. in 1998-99,[1,2] TiO$_2$-derived nanotubes obtained by a hydrothermal method have been attracted much attention because of their fascinating microstructure and promising photo electrochemical applications. The nanotubes can be prepared by a simple and environmentally innoxious processing. Several groups have tried to improve or to modify the yield and structure of the nanotubes.[3-6] The formation mechanism of nanotubes have been also discussed based on the TiO$_2$-anatase-structure model.[7,8] Some recent studies, however, suggest the as-prepared nanotubes by the hydrothermal method contained a fair amount of water, and that they should be recognized as trititanate H$_2$Ti$_3$O$_7$[9-13] or lepidocrocite titanate H$_x$Ti$_{2-x/4}$Y$_{x/4}$O$_4$ (x~0.7).[14] Table I summarizes the previous works on TiO$_2$-derived nanotubes synthesized by alkali treatment method. The present authors consider that such hydrated structure will offer another possibility to design various TiO$_2$-relating structures by post-treatments, such as well-controlled heat treatment.[15]

In this paper, heat treatment effect on the structure of TiO$_2$-derived nanotubes has been investigated by using high-temperature X-ray diffraction (HT-XRD), thermogravimetry-differential thermal analysis (TG-DTA) and *in situ* HT-SEM observation.

EXPERIMENTAL PROCEDURE

As shown in our recent work,[15] the nanotube preparation method was basically the same as described in previous papers.[2, 9-14] A commercial, fine TiO$_2$ (anatase) powder (Ishihara Sangyo Ltd., ST-01) was used as a starting material. In typical manner, 150 mg of TiO$_2$ powder was dispersed in NaOH aqueous solution (10 M, 50 ml), and put into a Teflon-lined stainless autoclave. The autoclave was heated at 110-150 °C for 72 h. After it was cooled to room temperature, the precipitated powder was washed with HCl aqueous solution and distilled water, and dried in a rather mild condition (60 °C for 12 h). Microstructure of the obtained powder was observed by transmission electron microscopy (TEM, JEOL JEM-200CX).

Table I TiO₂-derived nanotubes synthesized by hydrothermal method.

Reference	TiO₂ source (typical condition)	Alkali treatment conditions (typical condition)	Post treatment	Proposed Nanotubes' phase & composition*	Nanotubes' diameter	Nanotubes' length & surface area	Remarks by reference's authors
Kasuga et al. [1] (1998)	TiO₂-SiO₂ by sol-gel, or commercial P-25 (70% anatase-30% rutile)	110°C, 20h in 10 M NaOH (5 mg TiO₂/20mL.)	0.1M HCl and water washing	anatase, TiO_2 (EDS, SAED, XRD (pattern not shown))	Outer:~ 8 nm Inner:~ 5 nm	~ 100 nm ~ 400 m²/g	SiO_2 plays no role (mechanism was not given)
Kasuga et al. [2] (1999)	commercial rutile powder (JRC-TIO-3)	110°C, 20 h in 10 M NaOH (12.5g TiO₂/100mL.)	0.1M HCl and water washing	anatase, TiO_2 (containing H_2O)	Outer:~ 8 nm Inner:~ 5 nm	~ 100 nm 246 m²/g	HCl treatment is needed to change chemical bonds
Seo et al. [3] (2001)	anatase powder by precipitation (< 10 nm)	100-200°C, stirred 12 h in 5 M NaOH (simple refluxing)	vacuum filtering and washing	anatase, TiO_2 (TEM, SAED) not shown), Raman, TG-MS)	Inner : 7-8 nm (200C treated)	200-250 nm (200°C treated)	Higher temperature yielded more NTs.
Zhang et al. [4] (2002)	fine rutile powder (168 m²/g)	110°C, 20 h in 10 M NaOH	0.1M HNO₃ and water washing	Not accounted	Inner:~ 4 nm	~ 500 nm 320 m²/g	HNO₃ washing is better
Lin et al. [5] (2002)	anatase powder (10m²/g)	110°C, stirred 20-30 h in 10 M NaOH	0.1M HCl and water washing	anatase + rutile (trace), TiO_2 (EDS, SAED, XRD, ICP-AES)	Outer: 10-100nm Inner:~16 nm	2-18 micron 258 m²/g	Stirring during NaOH treatment yields long NTs.
Mao et al. [6] (2003)	commercial anatase powder (Aldrich)	Oil bath, 20 h in 10 M NaOH	0.1M HCl and water washing	anatase, TiO_2 (XRD)	Outer:~ 8 nm Inner: 4 nm	~ several hundreds nm	TiO₂ NTs can be precursor for perovskite NTs
Wang et al. [7] (2002)	anatase powder (100-150 nm)	Following Kasuga et al.[1]	Following Kasuga et al.[1]	anatase, TiO_2 (HRTEM, EELS, SAED)	Outer: 8~ 10nm Inner:~ 5nm	~ several hundreds nm	Scrolling mechanism. (3D→ 2D→ 1D)
Yao et al. [8] (2003)	TiO₂ (anatase) by hydrolysis	150°C, 12 h in 10 M NaOH	0.1M HNO₃ and water washing	anatase, TiO_2 (HRTEM, SAED, Raman)	Outer:~ 10nm Inner:~ 4~5 nm	~ several hundreds nm	lepidocrocite-related TiO₂-sheet as intermediate
Du & Chen et al. [9-12] (2001-2003)	TiO₂ by sol-gel, or commercial anatase	130°C, 24-72 h in 10 M NaOH	0.1M HCl and water washing	layered titanate, $H_2Ti_3O_7$ (HRTEM, EDS, EELS, SAED, XRD, IR)	Outer:~ 9 nm Inner:~ 4~5 nm	~ several hundreds nm 195 m²/g	Trittanate ($H_2Ti_3O_7$) model Scrolling mechanism.
Sun and Li [13] (2003)	TiO₂ by sol-gel (anatase)	100-180°C, >48 h in 10 M NaOH alcohol washing	water, acetone, or alcohol washing	layered titanate, $Na_xH_{2-x}Ti_3O_7$ (HRTEM, XRD, TG, DSC)	Various NTs were reported	Various NTs were reported	Na ion stabilizes NT structure
Ma et al. [14]	anatase powder	110-150°C, 12-72 h in 10 M NaOH	diluted HCl and water washing	layered titanate, lepidocrocite $H_xTi_{2-x/4}Y_{x/4}O_4$ (HRTEM, XRD, ED, TG)	Outer:~ 10 nm Inner: 5 nm	Not accounted	lepidocrocite model

*Methods for the determination or estimation of phase and composition.
EDS: Energy dispersive X-ray spectroscopy; EELS: Electron energy-loss spectroscopy; SAED: Selected-area electron diffraction.

Room temperature XRD patterns of the starting TiO_2 and the hydrothermally-treated powders were obtained by a RIGAKU RINT-2100 diffractometer ($CuK\alpha$, 40 kV and 40 mA). Effect of the heat-treatment on crystal structure of TiO_2-derived nanotubes was studied by HT-XRD analysis, using a RIGAKU RINT-2500 diffractometer with a platinum direct heating stage. A surrounding heater was added to extend the soaking zone, which allowed to put a sufficient amount of powder on the direct heating stage, and to improve S/N ratio. XRD patterns were acquired in the range between room temperature and 800 °C in air, after 3 min holding at each temperature (ramp rate: 10 °C/min). The heat-treatment effect was also studied via TG-DTA (10 °C/min, Shimadzu, DTG-50) in a static air atmosphere. *In situ* HT-SEM observation was preliminarily conducted using a field emission scanning electron microscope (SEM, Hitachi, S-5000) attached with tungsten-coil heating stage.[16,17]

RESULTS AND DISCUSSION

Fig. 1 shows a TEM micrograph (low magnification) of TiO_2-derived nanotubes; the insert indicates a selected area electron diffraction (SAED) pattern for group of nanotubes. A large quantity of rather long nanotubes (~ 1 μm in length) can be seen in the picture. There were some particle-like portions in the figure, but they were actually composed of kinks of nanotubes.[15] The nanotubes had a diameter of ~ 10 nm, which was in good agreement with previous studies.[2, 9-14]

XRD patterns obtained at room temperature are given in Fig. 2. The starting powder consisted of pure and fine anatase phase (Fig. 2(a)). In contrast, the as-synthesized nanotube powder had a different crystal structure (Fig. 2(b)). It can be assigned to trititanate $H_2Ti_3O_7$ as proposed by Peng's group,[9-12] but more preferably expressed as $H_2Ti_3O_7 \cdot nH_2O$.[15] The reflection peak at d~9.2 Å (0.92 nm) corresponds to the 200 reflection, which should be expanded by interlayer H_2O molecules (*see following*). Note that the 200 reflection of $H_2Ti_3O_7$ phase prepared by conventional solid-state reaction and ion-exchange method was at d=7.87 Å, reported by Feist et al.[18,19]

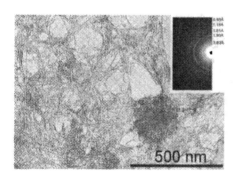

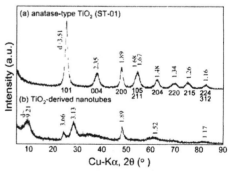

Fig. 1 Low magnification TEM image of TiO_2-derived nanotubes (insert: selected area diffraction image (SAED) for a wide area).

Fig. 2 XRD patterns of (a) starting material (anatase-type TiO_2) and b) hydrothermally-treated material (TiO_2-derived nanotubes).

Fig. 3(a) shows HT-XRD patterns for the as-synthesized nanotubes.[15] The patterns are plotted from $2\theta = 9.2°$ (d~9.6) to avoid the strong incident beam effect: a restriction of the RIGAKU-2500 system with PSPC-type detector. The XRD pattern obtained at room temperature (the bottom pattern in Fig. 3) is substantially identical to that in Fig 2 (b). At 200°C, clear peak shift (to $2\theta = 11.2°$) was observed. The reflection peak at $2\theta = 11.2°$ corresponds to the 200 reflection of $H_2Ti_3O_7$ phase, $d_{200} \sim 7.9$ Å. viz., the interlayer spacing of layered titanate.

The peak-shift behavior from room temperature to 200 °C can be well-explained by the dehydration of interlayer water (see Fig. 3 (b)). XRD pattern changed drastically at 700 °C, and further change was observed at 800 °C. At 800 °C, TiO$_2$(B)-like phase (a metastable polymorph of titanium dioxide[18, 20-23]) and rutile phase were identified. TiO$_2$ (B) was firstly reported by Marchand and Tournoux et al. in 1980.[20] Naturally occurring TiO$_2$ (B) was also identified in a natural anatase crystal by Banfield et al. in 1991.[23] Feist et al.[18] studied the formation mechanism of TiO$_2$ (B) from various layered titanates in detail. According to the Feist's report, observed phases at 700 °C (and some unindexed peaks at 800 °C) can be attributable to intermediate phases, such as $HTi_3O_{6.5}$ and $H_{0.5}Ti_3O_{6.25}$.

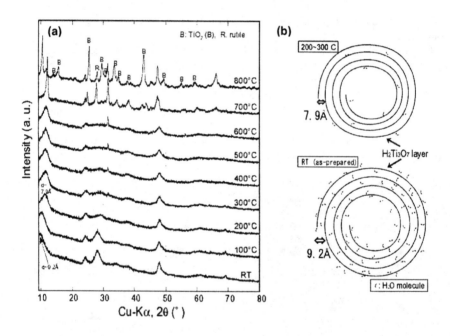

Fig. 3 (a) HT-XRD patterns for the as-prepared TiO$_2$-derived nanotubes, and (b) schematic illustration (cross section view) of TiO$_2$-derived nanotube: (lower) an as-prepared nanotube containing interlayer water with d_{200}~9.2 Å, (upper) a nanotube heated at 200-300 °C or in high-level vacuum (e.g., under TEM observation) with d_{200}~7.9 Å.

It is noteworthy that the nanotube's phase stability is strongly affected by the residual sodium ions, as pointed out by Sun et al.[13] In this study, sodium ions in the nanotubes were exchanged by protons via HCl treatment and H_2O washing process. However, a tiny amount of residual sodium ion (less than EDS-detection level) might be remained. If so, the unindexed peak at ~ 800 °C could be attributable to sodium hexatitanate, $Na_2Ti_6O_{13}$ (or other related sodium titanates). Therefore, further HT-XRD study in combination with detailed chemical analysis should be required in future.

TG-DTA curves of the as-prepared TiO_2-derived nanotubes are given in Fig. 4. The endothermic peak at 84.4 °C is attributable to the desorption of adsorbed water. At ~ 120° C, TG curve was slightly inflected (i.e., became gentle), which may correspond to the dehydration of interlayer water. Above 200 °C, weight loss became much gentler, suggesting the slow evaporation of structural water in $H_2Ti_3O_7$. These phenomena are in very good agreement with those of HT-XRD study. The exothermic peak at 709 °C is presumably attributable to the latent heat for the phase transformation from metastable $TiO_2(B)$ to stable rutile phase. From TG curve, the composition of as-prepared nanotubes (mildly dried and handled in normal air atmosphere) can be expressed as $\sim H_2Ti_3O_7 \cdot 3H_2O$. Recently, an alternative model for TiO_2-derived nanotubes, i.e., lepidocrocite titanate $H_xTi_{2-x/4}\Upsilon_{x/4}O_4$ (x~ 0.7), has proposed by Ma et al.,[14] since the reflection at d~9.2 Å in XRD pattern is difficult to be explained by the $H_2Ti_3O_7$ model, and TiO_2 (B) was not identified in their study after 500 °C annealing for the nanotubes (heating duration was not given[14]). The present HT-XRD work, however, demonstrated that the reflection at d~9.2 Å can be well-explained by the swelling via water intercalation into 200 plane of $H_2Ti_3O_7$ (Fig. 3(b)). In addition, TiO_2 (B) (dehydration product of $H_2Ti_3O_7$[18]) was actually identified during the heating process (Fig. 3).

Fig. 5 comprehensively illustrates the nanotube formation process, post-heat treatment process, and earlier work on titanate fiber processing by Watanebe et al.[24,25] A very recent work,

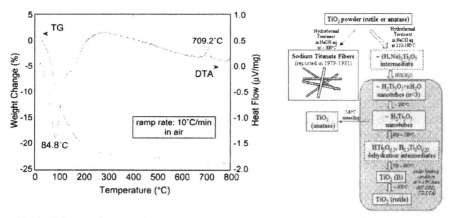

Fig.4 TG-DTA diagrams for the as-prepared TiO_2-derived nanotubes.

Fig. 5 Nanotube formation process, post-heat treatment process, and earlier work on titanate fiber processing.

including a series of TEM observation for intermediate products during the hydrothermal process and *ab initio* calculations, has given a new insight into the nanotube formation mechanism,[12] corresponding to the upper right part of Fig. 5.

To clarify the microstructural change during the heat-treatment, preliminarily *in situ* high-temperature SEM observation has been carried out.[26] Fig. 6 (a) and (b) show SEM micrographs of the TiO_2-derived nanotubes before and after heating, respectively. Before heating, nanotube bundles and separated nanotubes were observed (Fig. 6 (a)). With heating, nanotube bundles and separated nanotubes changed into one-dimensionally connected nano particles (Fig. 6 (b)). This direct observation well explains the fact that the nanotube structure of the hydrothermally prepared TiO_2-derived nanotubes became unstable around 500 °C, and changed into anatase-type particles or dense fibers.[14,27] This HT-SEM study suggests a future possibility that the hydrothermally prepared TiO_2-derived nanotubes can be used as precursor for one-dimensional nanostructured TiO_2. Such materials might be useful for photoelectric applications with rapid electron transfer, e.g. for photocatalyst or dye-sensitized solar cells. Detailed *in situ* HT-TEM will be required to reveal the microstructural change in atomic scale.

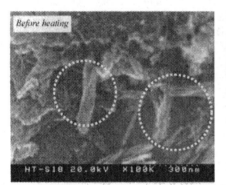 

Fig. 6 SEM micrographs of the TiO_2-derived nanotubes before and after heating, respectively (*in situ* heating observation).

CONCLUSIONS

In this study, TiO_2-derived nanotubes were prepared by hydrothermal treatment of TiO_2 powder in NaOH aqueous solution. HT-XRD and TG-DTA studies revealed the dehydration and the phase transformation behavior of nanotubes. The constituent phase of the as-prepared nanotube can be assigned most probably as $H_2Ti_3O_7 \cdot nH_2O$ (n < 3). Post-treatment for nanotubes enabled to produce TiO_2-related metastable materials (*e.g.* TiO_2 (B)), and such metastable materials are also promising for photo electrochemical and catalytic applications like titanate nanotubes. The hydrothermally prepared TiO_2-derived nanotubes can be used as precursor for one-dimensional nanostructured TiO_2, as predicted by the preliminary *in situ* SEM study.

ACKNOWLEDGMENTS

The authors wish to thank Mr. T. Nokuo (JEOL Ltd.) and Ms. Ying Hu (AIST) for the experimental help. We acknowledge Dr. Kanzaki (AIST), Prof. Isoda, Prof. Kurata and Prof. Yoko (Kyoto University), and Prof. Niihara and Prof. Sekino (Osaka University) for the use of apparatuses. A part of this work has been supported by 21COE program "Establishment of COE on Sustainable Energy System", and "Nanotechnology Support Project" of the Ministry of Education, Culture, Sports, Science and Technology (MEXT), Japan. Travel expense for the 106th Annual Meeting of ACerS has been supported by Kansai Research Foundation for technology promotion (KRF).

REFERENCES

[1]T. Kasuga, M. Hiramatsu, A. Hoson, T. Sekino and K. Niihara, "Formation of Titanium Oxide Nanotube," *Langmuir*, **14** [12] 3160-3163 (1998).

[2]T. Kasuga, M. Hiramatsu, A. Hoson, T. Sekino and K. Niihara, "Titania Nanotubes Prepared by Chemical Processing," *Adv. Mater.*, **11** [15] 1307-1311 (1999).

[3]D.-S. Seo, J.-K. Lee and H. Kim, "Preparation of Nanotube-Shaped TiO_2 Powder," *J. Cryst. Growth*, **229** [1] 428-432 (2001).

[4]Q. H Zhang, L. A Gao, J . Sun and S. Zheng, "Preparation of Long TiO_2 Nanotubes from Ultrafine Rutile Nanocrystals," *Chem. Lett.*, **31** [2] 226-227 (2002).

[5]C. H. Lin, S. H. Chien, J. H. Chao, C. Y. Sheu, Y. C. Cheng, Y. J. Huang, C. H. Tsai, "The Synthesis of Sulfated Titanium Oxide Nanotubes," *Cataly. Lett.*, **80** [3-4] 153-159 (2002).

[6]Y. B.Mao, S. Banerjee, and S. S. Wong, "Hydrothermal Synthesis of Perovskite Nanotubes," *Chem Comm.*, [3] 408-409 (2003).

[7]Y. Q. Wang, G.. Q. Hu, X. F. Duan, H. L. Sun and Q. K. Xue, "Microstructure and Formation Mechanism of Titanium Dioxide Nanotubes," *Chem. Phys. Lett.*, **365** [5-6] 427-431 (2002).

[8]B. D. Yao, Y. F. Chan, X. Y. Zhang, W. F. Zhang, Z. Y. Yang, and N. Wang, "Formation Mechanism of TiO_2 Nanotubes," *Appl. Phys. Lett.*, **82** [2] 281-283 (2003).

[9]G. H. Du, Q. Chen, R. C. Che, Z. Y. Yuan, L. M. Peng, "Preparation and Structure Analysis of Titanium Oxide Nanotubes," *Appl. Phys. Lett.*, **79** [22] 3702-3704 (2001).

[10]Q. Chen, G. H. Du, S. Zhang and L. M. Peng, "The Structure of Trititanate Nanotubes,"*Acta Crystallogr. B*, **58** [4] 587-593 (2002).

[11]Q. Chen, W.Z. Zhou, G. H. Du, L. M. Peng, "Trititanate Nanotubes Made via a Single Alkali Treatment," *Adv. Mater.*, **14** [17] 1208-1211 (2002).

[12]S. Zhang, L. M. Peng, Q. Chen, G. H. Du, G. Dawson and W. Z. Zhou,"Formation Mechanism of $H_2Ti_3O_7$ Nanotubes," *Phys. Rev. Lett.*, **91** [25] 256103 (2003).

[13]X. Sun, and Y. Li, "Synthesis and Characterization of Ion-Exchangeable Titanate Nanotubes," *Chem. Euro. J.*, **9** [10] 2229-2238 (2003).

[14]R. Z. Ma, Y. Bando, and T. Sasaki, "Nanotubes of Lepidocrocite Titanates," *Chem. Phys Lett.*, **380** [5-6] 577-582 (2003).

[15]Y. Suzuki and S. Yoshikawa, "Synthesis and Thermal Analyses of TiO_2-Derived Nanotubes Prepared by the Hydrothermal Method," *J. Mater. Res.*, **19** [4] 982-985 (2004).

[16]Y. Suzuki, P.E.D. Morgan, T. Sekino and K. Niihara, "Manufacturing Nano-Diphasic Materials from Natural Dolomite - *In-situ* Observation on Nano-Phase Formation Behavior -," *J.*

Am. Ceram. Soc., **80** [11] 2949-2953 (1997).

[17]Y. Suzuki, T. Sekino, T. Hamasaki, K. Ishizaki and K. Niihara, "*In Situ* Observation of Discrete Glassy SiO_2 Formation and Quantitative Evaluation of Glassy SiO_2 in $MoSi_2$ Compacts," *Mater. Lett.*, **37** [3] 143-148 (1998).

[18]T. P. Feist and P. K. Davies, "The Soft Chemical Synthesis of TiO_2 (B) from Layered Titanates," *J. Solid State Chem.*, **101**, 275-295 (1992).

[19]ICDD-JCPDS Powder diffraction file, #47-0561, 1997.

[20]R. Marchand, L. Brohan and M. Tournoux, "TiO_2 (B) A New Form of Titanium Dioxide and the Potassium Octatitanate $K_2Ti_8O_{17}$," *Mater. Res. Bull.*, **15**, 1129-1133 (1980).

[21]L. Brohan, A. Verbaere, M. Tournoux, "La Transformation TiO_2 (B) $\rightarrow$ Anatase," (in French), *Mater. Res. Bull.*, **17**, 355-361 (1982).

[22]ICDD-JCPDS Powder diffraction file, #35-0088, 1985.

[23]J. F. Banfield, D. R. Veblen, and D. J. Smith, "The Identification of Naturally Occurring TiO_2 (B) by Structure Determination Using High-Resolution Electron Microscopy, Image Simulation, and Distance-Least-Squares Refinement," *Am. Mineral.*, **76**, 343-353 (1991).

[24]W. Watanabe, Y. Bando and M. Tsutsumi, "A New Member of Sodium Titanates, $Na_2Ti_9O_{19}$", *J. Solid State Chem.*, **28**, 397 (1979).

[25]W. Watanabe, "The Investigation of Sodium Titanates by the Hydrothermal Reactions of TiO_2 with NaOH," *J. Solid State Chem.*, **36**, 91 (1981).

[26]Y. Suzuki, R. Yoshida, T. Sekino, and S. Yoshikawa, "*In Situ* Observations of Mor- phological Change during Heating for TiO_2-Derived Nanotubes Prepared by Hydrothermal Method," *Proceedings of 8th Asia-Pacific Conference on Electron Microscopy (8APEM)*, June 7 to11, 2004 (Kanazawa, Japan), Japanese Society of Microscopy, in press.

[27]T. Sekino, "Does the One-Dimensional Nanospace in Titania Nanotubes contribute to the Functionalization ?" (in Jpn.), *Preprints of the 6th Kansai Branch Forum for Young Scientists and Engineers on Ceramic Studies*, The Ceramic Society of Japan, 2003, p28.

PREPARATION OF TITANATE NANO-ROD ARRAY ON TITANIUM SUBSTRATES BY NOVEL MICROFLUX METHOD

Yongxing Liu, Kazuya Okamoto, Satoshi Hayakawa, Kanji Tsuru and Akiyoshi Osaka
Biomaterials Laboratory, Faculty of Engineering
Okayama University
3-1-1, Tsushima-naka, Okayama, 700-8530, Japan

ABSTRACT

We proposed a new approach on preparing titania or titanate nano- or micro-rod arrays on metallic titanium (α-Ti) surfaces by coating a layer of sodium tetraborate or potassium metaborate on titanium substrates and subsequent thermal treatment. Thin-film X-ray diffraction analysis indicated that the sodium tetraborate gave rutile (TiO_2: PDF# 21-1276) micro-rod array. The potassium metaborate, in contrast, yielded a potassium titanate ($K_2Ti_6O_{13}$: PDF# 40-0403) nano-rod array. The rods in the arrays grew almost perpendicular to the substrate surface.

INTRODUCTION

One-dimensional nanostructures of polymers, metals and semiconductors in forms of rods, tubes and others, have attracted much attention for a broad range of potential applications, including catalysts, electronics, sensors, photonics, micromechanical devices and others. However, to obtain scale-up functional devices with highly ordered nanorod or nanotube arrays is essential. A number of fabrication techniques have been employed or challenged to fabricate highly ordered nanorod or nanotube arrays. For instance, highly ordered arrays of anisotropic nanorods of ZnO have been successfully created via various methods.[1-3] Many of these fabrication processes were also applicable for other materials, such as TiO_2, but it is complicated because of the templates used or the chemical process involved. TiO_2 and the potassium titanate are all semiconductors. Potassium titanates and their derivatives, which were known for their photocatalytic activity,[4] ion-exchange reaction property,[5] high chemical and thermal stability,[6] high insulating ability were usually prepared by hydrothermal method, or sintering, melting, flux melt from K_2O-TiO_2 system at the temperature over 1000 °C.[7]

In this study we proposed a simple technique to prepare nano-scale rod arrays of rutile type

TiO$_2$ or a potassium titanate on metallic titanium substrates, in which powders of sodium tetraborate glass or potassium metaborate were used for coating, followed by heat treatment at 700 °C for 5h. Both as-achieved structures of the rods were characterized in terms of nano scale in diameter.

EXPERIMENTAL PROCEDURE

Reagent grade sodium or potassium carbonates and boron oxides (Nacalai Tesque Inc., Kyoto, Japan) were used to prepare the batches of Na$_2$O·2B$_2$O$_3$ and K$_2$O·B$_2$O$_3$ composition. They were melted in a platinum crucible with a cover and subsequently placed in an electric furnace at 1200 °C for 1 h. The melts were then poured on a sheet of steel plate. The quenched glass was grounded into powders that were less than 45 μm in diameter by a planetary micro-mill pulverizer, pulverisette-7 (Fritsch, Germany) at speed of 600 rpm for 10 min.

A sheet of commercial pure titanium (Nilaco, Osaka, Japan) was cut to give pieces of Ti substrate with of 10 × 10 × 0.1 mm in size. The micro-scale glass powders obtained above were placed directly onto Ti substrates in the thickness of about 0.5 mm as shown in Fig.1. Then, the samples were heated up to 700 °C at the rate of 10 °C/min in an electric furnace, and maintained for 5 h. After cooling down to room temperature, the glass-coated samples were subsequently immersed in 80 °C water and kept for 5 h to remove the coating. Microstructures of the samples were observed after coating 30 nm gold by a scanning electron microscopy (JSM-6300, JEOL, Japan), which was operated under 20 kV acceleration voltage and 300 mA emission current. Crystal phases present in the samples were identified by thin-film X-ray diffraction (TF-XRD: CuKα) patterns taken by an X-ray diffractometer (RINT2500, Rigaku, Tokyo, Japan) operated at 40kV-200mA and at a scanning step of 0.1 °/s.

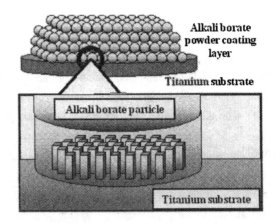

Fig.1 Schematic model of nano- or micro- rod array formation process (denoted as "Microflux method") by thermal treatment.

RESULTS

Fig. 2 and 3 show the scanning electron microscope image and the TF-XRD patterns for the microstructure on titanium substrates obtained by the coating of sodium tetraborate glass and subsequent removal with hot water treatment, respectively. After the glass coating was dissolved away in the hot water, the rod array structure appeared in discrete regions within the range of dozens of micrometers in length and width on the titanium surface. The array showed an array structure in which each rod is almost perpendicular to the substrate surface, as seen in Fig. 2. The rods had a rectangular bottom shape with dozens of nanometers in width and, while the length was estimated to be a few micrometers.

Fig. 2 SEM image of the micro-scale rod arrays on titanium substrate.

TF-XRD patterns in Fig. 3 gave sharp and strong diffraction peaks at 27.48° and 36.08° which were assigned to (110) and (101) planes of rutile (PDF# 21-1276), respectively. The peaks at 38° and 40° were assigned to metallic α-Ti. The EDX analysis (not presented here) detected only Ti and oxygen elements with the rough ratio of 1:2. This indicated that the sodium tetraborate glass coating yielded micro-scale rods of rutile (TiO_2) on the titanium substrate.

When potassium metaborate ($K_2O \cdot B_2O_3$) was used instead of

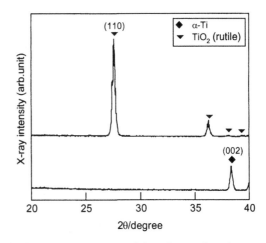

Fig. 3 TF-XRD patterns of the micro-scale rod arrays on titanium substrate and as-received titanium substrate.

sodium tetraborate glass, the highly ordered rod array was observed on the whole surface, as shown in Fig. 4. The arrays had well-ordered structures while the rods pointing upwards against titanium substrates. Note that the rods displayed a marked monodispersion in terms of width and length that were about 100 nanometers and several micrometers, respectively, which were estimated by the SEM images.

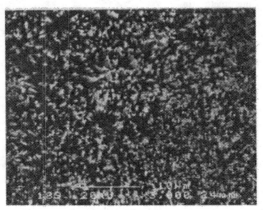

Fig. 4   SEM image of the nano-scale rod arrays on titanium substrate.
(Bar 1.0 μm)

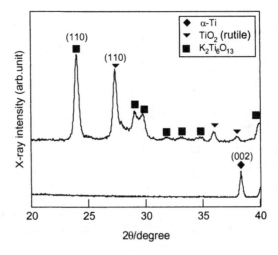

Fig. 5   TF-XRD patterns of the nano-scale rod arrays on titanium substrate and as-received titanium substrate.

The TF-XRD patterns of the nano-scale rod arrays on titanium substrates and as-received titanium substrate were shown in Fig. 5. The peaks at 24.10°, 29.2° and 29.8° were assigned to (110), (310) and (31-1) planes of $K_2Ti_6O_{13}$ (PDF# 40-0403), respectively. The peaks at 27.48° and 36.0° were assigned to (110) and (101) planes of rutile, respectively.

DISCUSSION

The nano- or micro-scale rod arrays of rutile and potassium titanates can be obtained on titanium surfaces after the heat treatment when the alkali-borate glass coatings were applied to the Ti substrates. Noted that the diffraction of rutile (110) plane and the diffraction of α-Ti (002) plane are predominant in Fig. 2. We can consider the topotaxy between the α-Ti (002) plane and rutile (110) plane as shown in Fig. 6. We notice a good correlation between the Ti-Ti atomic distances in α-Ti (002) plane; 2.9506Å and rutile (110) plane; 2.9587Å along c-axis.

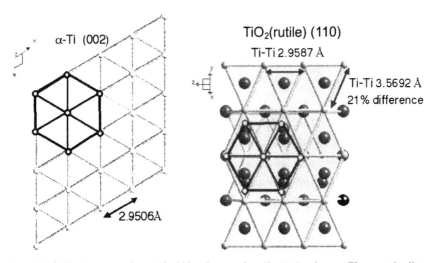

Fig. 6 Relation between the α-Ti (002) plane and rutile (110) plane. The atomic distances between Titanium atoms are also shown.

The topotaxy will dominate the crystal growth of TiO_2 (rutile) layers, irrespective of sodium tetraborate glass or potassium metaborate coatings.

The chemical reaction and crystal rods growth processes may involve five key points: 1) nano-crystalline TiO_2 (rutile) formed by thermal oxidation of α-Ti (002) plane in the oxygen environment, on the basis of topotaxy described in Fig. 6; 2) nano-crystalline TiO_2 (rutile) were attacked and eroded by alkali metal oxide component of alkali borates, and the erosion may be dependent on the basicities; 3) the as-formed TiO^{x-} species transported and then combined with alkali metal oxides such as Na_2O and K_2O forming titanates, all of which were thermodynamically stable under these given conditions; 4) the existence of B_2O_3 in the glass may play the role of slowing down not only the chemical reaction process but also the crystal growth by the diluting effect on alkali metal oxides; 5) titanates $M_xTi_yO_z$ or TiO_2 crystallized on the titanium surface and grew up competitively over the titanium substrate surface as the complex process of erosion, transportation, combination, crystallization and crystal growth. The reactivity of sodium tetraborate glass and potassium metaborate to titanium was supposed different at the temperature of 700 °C because of different melting temperatures and the other chemical discrepancy. Since the sodium tetraborate glass may have lower reactivity in forming sodium titanates, or the solubility of the sodium tetraborate in hot water may be higher than that of potassium metaborate, TiO_2 rutile rods grew up predominantly as shown in Fig. 2. As for the potassium metaborate, it could react actively with titanium at 700 °C so the potassium titanates rods grew up predominantly and quickly, thus the as-formed titanate rod array displayed a uniform and well-ordered structure.

CONCLUSION

TiO_2 (rutile) and $K_2Ti_6O_{13}$ in the form of rod arrays were fabricated on titanium substrates by a simple interface reaction between glassy coatings and metallic titanium. The as-achieved rod arrays showed well-ordered structures. Combining the high quality in products and the easiness in fabrication, this novel technique promises us one simple method to prepare highly ordered nano-scale arrays of various functional materials on various substrates.

ACKNOWLEDGEMENTS

This study is financially supported by The Sumitomo Foundation and The Sanyo Broadcasting Foundation.

REFERENCES

[1]L. Vayssieres, "Growth of arrayed nanorods and nanowires of ZnO from aqueous solutions," *Adv. Mater.*, **15** [5] 464-6 (2003).

[2]M. H. Huang, Y. Wu, H. Feick, N. Tran, E. Weber, and P. Yang, "Catalytic Growth of Zinc Oxide Nanowires by Vapor Transport," *Adv. Mater.*, **13** [2] 113-6 (2001).

[3]J. H. Choi, H. Tabata and T. Kawa. "Initial preferred growth in zinc oxide thin films on Si and amorphous substrates by a pulsed laser deposition," *J. Cryst. Growth*, **226** [4] 493-500 (2001).

[4]K. Sayama, H. Arakawa, "Effect of Na_2CO_3 addition on photocatalytic decomposition of liquid water over various semiconductor catalysts," *J. Photochem. Photobiol. A: Chem.*, **77** [2-3] 243-247 (1994).

[5]T. Sasaki, Y. Komatsu, and Y. Fujiki, "Rb^+ and Cs^+ incorporation mechanism and hydrate structures of layered hydrous titanium dioxide," *Inorg. Chem.*, **28**, 2776-2779 (1989).

[6]C.-T. Lee, M.-H. Um, H. Kumazawa, "Synthesis of Titanate Derivatives Using Ion-Exhange Reaction," *J. Am. Ceram. Soc.*, **83** [5] 1098-1102 (2000).

[7]G. L. Li, G. H. Wang, and J. M. Hong, "Synthesis and characterization of $K_2Ti_6O_{13}$ whiskers with diameter on nanometer scale," *J. Mater. Sci. Lett.*, **18**, 1865-1867 (1999).

CONTROLLING THE STRUCTURE OF ALIGNED CARBON NANOTUBES ON SILICON-CARBIDE WAFERS

Michiko Kusunoki, Toshiyuki Suzuki, Chizuru Honjo and Tsukasa Hirayama
Japan Fine Ceramic Center
2-1-4 Mutsuno, Atsuta, Nagoya 456-8587
Japan

ABSTRACT
　　Well-aligned carbon nanotube (CNT) films were synthesized by surface decomposition of silicon carbide (SiC) (0001). In the initial stage of the decomposition, at 1200-1250 °C, the generation of semispherical carbon caps of several nanometers all over the surface of SiC was found by cross-sectional transmission electron microscopy (TEM) and atomic force microscopy (AFM). The diameter of the grown CNTs is determined initially by that of the nanocaps. Moreover, zigzag-type CNTs are selectively produced by surface decomposition of a well-polished SiC single crystal. The SiC wafer was heated to 1500 °C at a very small heating rate under vacuum. TEM and electron diffraction patterns revealed that almost all the well-aligned CNTs formed perpendicularly to the SiC (0001) surface are double-walled, zigzag type. In addition, the results of high-resolution electron microscopy (HREM) indicate that the zigzag-type structure evolves from Si-C hexagonal networks in the SiC crystal by the collapse of carbon layers remaining after the process of decomposition.

INTRODUCTION
　　Since the discovery of carbon nanotubes (CNTs), their unique properties and remarkable potentials have surprised many people and attracted much interest from a great number of researchers. The CNT is presently one of the most anticipated new materials not only in scientific but industrial fields. So far, various kinds of synthetic methods such as arc-discharged,[1, 2] laser ablation,[3] and chemical vapor deposition (CVD) method[4-6] have been reported. Especially, CVD has been improved to produce single-walled carbon nanotubes (SWNTs) with both of high quality and large quantity. To employ the CNTs with reliance as an industrial material, it is important to control the structure of CNTs, diameter, length, and chirality. It becomes possible to a certain degree to change distribution of the diameter of CNTs and selectively synthesize SWNT/MWNTs by controlling the particle diameter, kind of metal catalysts or the gas and substrate temperature.[7-9] Contrarily, it is not easy to control the length of the CNTs since the growth rate is so high.
　　Moreover, the previous methods, usually result in three types, armchair, zigzag, and chiral-type CNTs, co-existing in the final product. And it has been confirmed theoretically[10-12] and experimentally[3,13,14] that armchair-type SWNTs are always be metallic, while zigzag- and chiral-type CNTs are alternative between metallic and semiconducting states as a function of the chiral vector. An important requirement for fundamental research and industrial applications is that all CNTs for a certain application be produced with the same structure. However, there is no effective means of controlling the type of CNT structures produced. Accordingly, the first step in attaining this control is to understand how the chirality is determined.
　　The authors previous work showed that aligned CNTs are self-organized by surface decomposition of the SiC (0001) C-face,[15] and we also identified the formation mechanism of the CNTs.[16,17] In regards to this method, silicon atoms are evaporated selectively from the surface, and the residual carbon atoms are recrystallized as aligned CNTs. The current paper reports on how the CNT films are constructed and how the structure of the aligned CNTs synthesized by this method can be restricted to the zigzag type.[18] A formation mechanism based on high-resolution electron microscopy

(HREM) observations of the interface between the SiC substrate and the CNTs is proposed.

EXPERIMENTAL

A 6H-SiC single crystal wafer, 0.25 mm thick with well-polished $(000\bar{1})$ C-faces (Ra<0.38 nm) on axis and a nitrogen dopant concentration of about 10^{18} cm^{-3}, were prepared as samples. They were cut into 1×4 mm^2 sections along the $(1\bar{1}00)$ and $(11\bar{2}0)$ planes and ultrasonically cleaned in dichloromethane, acetone, and ethanol, successively, each for 10 min. Some of the sections were heated to 1700 °C in a vacuum of 1×10^{-2} Pa at a rate of 20 °C/min, and kept at 1700 °C for half an hour. The others were heated to 1500 °C at a rate of 1 °C/min in the same vacuum, and kept at 1500 °C for 10 h. Moreover, the $(1\bar{1}00)$ and $(11\bar{2}0)$ surfaces, the narrow side faces of the other SiC sections were polished. These samples were also heated under the same conditions. The samples were thinned by a conventional ion-thinning method and observed from the plan-view and cross-sectional directions by electron microscopes, Topcon 002B and JEOL 2010 TEM, at an acceleration voltage of 200 kV.

RESULTS AND DISCUSSION
An Aligned CNTs Formed on SiC(000$\bar{1}$) Surface

Figure 1 is a TEM image (a) and an electron-diffraction pattern (b) obtained from the C-face of the SiC single-crystal wafer heated to 1700 °C at a rate of 20 °C/min and kept at that temperature for half an hour. Well-aligned, straight, and dense CNTs (0.25 μm in length) perpendicular to the SiC surface can be seen. Figure 2 shows the HREM micrograph of the tip of the CNTs, three to five-layered and 3-5 nm in diameter.

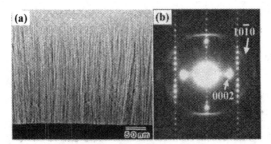

Fig. 1 TEM image and the electron diffraction pattern of an aligned carbon nanotube film on a SiC (000$\bar{1}$) C-face formed by heating to 1700 °C at a heating rate of 20 °C/min.

Fig. 2 HREM image of the tip of a CNT film. The amorphous contrast on the upper side is due to the glue used for TEM sample preparation.

To investigate the mechanism of surface decomposition of SiC, a surface heated at lower temperature was observed by TEM and atomic force microscopy (AFM). Figure 3 shows a micrograph of the SiC (0001) C-face heated at 1250 °C for half an hour. Semispherical particles of four-layered graphite with outer diameter of 5 nm and height of 1-2 nm are clearly dispersed as shown by the arrows in Figure 3(a). The density of these graphite particles was confirmed to increase with increasing temperature. Figure 3(b) is an AFM image of the (0001) surface heated at 1200 °C for one hour. Several-nanometer-size dots can be clearly seen. They were observed uniformly all over the surface of the SiC wafer. These results explain the initial stage of the decomposition of SiC and formation of CNTs. That is, nanocaps are formed in the initial stage of the formation of the CNTs as shown schematically in Figure 3(c).

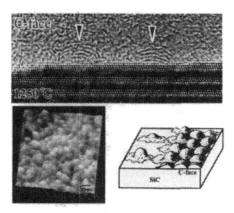

Fig. 3. Formation of nanocaps in the initial stage of the surface decomposition.

Once the nanocaps have formed on the surface, graphene sheets grow cylindrically with the diameter of each nanocap toward the inner-side of the SiC, thus eroding the SiC perpendicularly.

Controlling the Structure of CNTs

Figure 4(a) is a TEM micrograph of an SiC (0001) C-face heated to 1500 °C at a rate of 1°C/min and kept at that temperature for 10 h in a vacuum of 10^{-2} Pa. Aligned CNTs (225 nm long), similar to those in the micrograph shown in Figure 1 (a), can be seen. The micrograph taken along the plan-view direction is shown on the left of Figure 4(a). The structure was confirmed to consist of mainly double-walled CNTs (DWNTs). According to measurements of the diameters of over 200 tube rings in the micrographs taken along the plan-view direction, the average diameter of the DWNTs was about 3 ± 1 nm. The image contrast of CNTs in Figure 4(a) seems finer and denser than that in Figure 1(a), because the diameter of the CNTs shown in Figure 4(a) is smaller.

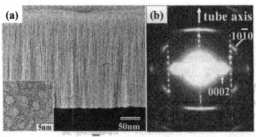

Fig. 4 Aligned CNT film on the SiC (0001) C-face formed by heating to 1500 °C at a heating rate of 1°C/min and kept at 1500 °C. (a) TEM micrographs taken from cross-sectional and plan-view directions (lower left). (b) The selected-area electron diffraction pattern of the CNTs.

Figure 4(b) shows a selected-area electron diffraction pattern obtained from the cross-sectional CNT film on the SiC shown in Figure 4(a). The diffraction pattern contains three kinds of reflections: six-fold $10\bar{1}0*$ spots from horizontal graphene net planes, 0002* and 0004* spots from vertical planes of the aligned CNTs, and net-pattern spots from the SiC crystal where the incident beam direction is parallel to $[11\bar{2}0]_{SiC}$ direction. The orientation relationship between the 0002* and $10\bar{1}0*$ spots in this pattern clearly shows that most of the CNTs have a zigzag-type structure.

Only the $10\bar{1}0*$ reflections perpendicular to the 0002* reflection shown by arrows in Figure 4(b) are especially strong and elongated in the c* direction. The intensity consists of two types of reflections from the horizontal and vertical planes of the aligned CNTs. The streaked reflections and the 0002* correspond to the ones reflected from the vertical planes of the CNTs, where the incident-beam direction is parallel to the $[01\bar{1}0]$ of graphite. Since the 0002* spot shows no streaking, the 10-10* streak cannot be due to the shape of the crystal. Instead, this streak can be ascribed to the random shift of each SWNT within an MWNT along the CNT axis. The vertical planes of period $d_{10\text{-}10}$ are stacked independently, without correlation to each other, but are separated by the same distance, 0.34 nm (about c/2).

The wall of the zigzag-type CNTs is assumed to shift not only along the CNT axis but also about the CNT axis non-helically. Even in this case, the diffraction-intensity distribution from the vertical plane does not change, though the intensity is weakened by the shift and rotation. Furthermore, most of the adjacent DWNTs at a graphite-layer distance of about 0.344 nm would also be shifted relative to each other. Such a shift also causes the streak of the $10\bar{1}0*$ reflection. As a result, the intensity distribution of the electron diffraction pattern in Figure 4(b) is assigned to the shift of the zigzag-type CNTs along and around the CNT axis.

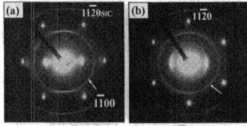

Fig. 5 Electron diffraction patterns of CNTs formed on (a) the $(1\bar{1}00)$ and (b) the $(1\bar{1}\bar{2}0)$ SiC surface planes.

Furthermore, to investigate the orientation relationship between the structures of the SiC crystal and the CNTs, SiC surfaces, except the (0001) planes, were heat-treated under the same conditions as those in the case shown in Figure 4. Figure 5 shows the electron-diffraction patterns of the CNTs formed on the polished 6H-SiC (1100) and (1120) planes, which are perpendicular to the (0001) plane. The irradiated beam directions are both parallel to the [0001] direction. In both of the electron-diffraction patterns, the 1100* graphite reflection for a pair of 0002* spots shows a Debye ring. This means that the CNTs formed on the SiC (1100) and (1120) surfaces at least have not constant chirality. Even on the SiC (0001) plane, the 1100* graphite reflection was arced when the sample was heated at 20°C/min as shown in Fig. 1(a). The above results strongly suggest that the atomic arrangement in the SiC crystal around the axis in the [0001] direction favors the formation of the zigzag-type CNT structure.

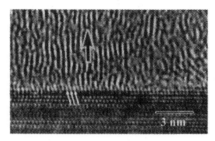

Fig. 6 HREM micrograph of the interface between MWNTs and the SiC single crystal observed along [1120] SiC.

Figure 6 shows a cross-sectional HREM micrograph of the interface between the CNTs and the SiC (0001) C-face heated at 1700 °C for 0.5 h. The (0002) graphite lattice fringes corresponding to the vertical planes of the CNT walls connect directly with the SiC crystal without forming any amorphous layer at the interface. Though the graphite lattice is almost parallel to the SiC (1010) plane perpendicular to the surface, at the boundary between them, they are a little tilted from the perpendicular direction against the surface. The graphite lattices ($d_{0002g} = 0.34$ nm) are roughly

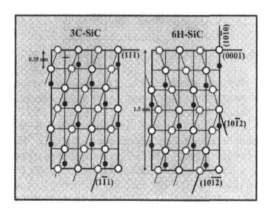

Fig. 7. Schematic of projected structures of 3C- and 6H-SiC along the $[10\bar{1}]_{3C}$ and $[\bar{1}2\bar{1}0]_{6H}$ directions, respectively.

matched and aligned with the ($10\bar{1}2$) or ($10\bar{1}2$) SiC planes (d_{10-12}=0.308 nm) in the last three layers of the twin structure responsible for the 6H-SiC stacking sequence, as shown in Figure 7. The {$10\bar{1}2$} planes of 6H-SiC (hexagonal-type) correspond to the {111} planes of 3C-SiC (fcc type), which have the highest carbon atomic density in the fcc crystal. The tubular graphite lattices are matched and aligned six times with the six-fold {$10\bar{1}0$} planes around the [0001] axis. This indicates that just after decomposition, the CNT walls inherit the atomic structure of the original SiC, and the wall planes are composed roughly of the planes with the highest carbon atom density in the SiC crystal.

Fig. 8. Schematic representations of one of Si-C hexagons in a 3C-SiC unit cell.

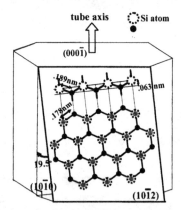

Fig. 9. A formation mechanism for a zigzag-type CNT along the c axis.

The {111} planes are composed of networks of slightly wavy hexagons consisting of three silicon atoms and three carbon atoms as shown in Figure 8. Accordingly, a monolayer network structure of these Si-C hexagons also exists on the ($10\bar{1}2$) plane as shown in Figure 9. With the outward diffusion of silicon atoms and the collapse of carbon atoms from the inner side of the CNT wall to the silicon positions, a carbon honeycomb structure with a zigzag edge perpendicular to the CNT axis is formed as shown in Figure 9. The Si-C distance projected on the ($10\bar{1}2$) plane is 0.178 nm and the network has to shrink at a rate of 20% to form graphene with C-C bonds of 0.142 nm. Since the ($10\bar{1}2$) plane is tilted by ±19.5° with respect to the ($10\bar{1}0$) plane along the CNT axis, this shrinkage along the CNT axis might favor the carbon network formed on the {$10\bar{1}2$} planes becoming parallel to the CNT axis. In this case the shrinkage rate is 6%.

It was shown in our previous papers[16,17] that nanocaps are generated on the SiC (000$\bar{1}$) surface at the initial stage, and then the diameter of the CNTs are determined by the diameter of the nanocaps.

As a result, following the initial formation of the caps, a zigzag-type CNT wall grows cylindrically along six-fold {1010} SiC planes around the c-axis. At this time, the shrinkage in the direction perpendicular to the tube axis can be reduced by about 10% by curving the graphene sheet to form a tube touching the hexagonal column of six {1010} planes internally. The high temperature used has enabled such a rearrangement.

CONCLUSION

A unique synthetic method, SiC surface decomposition, for forming well-aligned CNT films on SiC single-crystal wafers, was developed. TEM and AFM observations revealed that in the initial stage of the decomposition, carbon nanocaps with diameters of several nanometers on the SiC surface are generated. The diameter of the grown CNTs is considered to be determined by the size of the nanocaps.

Furthermore, it was shown that the present method can selectively produce zigzag-type CNTs, Accordingly, aligned, straight, and identically structured CNTs were produced for the first time by this method. The formation mechanism can be explained in terms of a topotaxial orientation relationship between graphene formed by SiC decomposition and the SiC crystal substrate. The present study indicates it is possible to control the chirality of CNTs by crystallographically restricting the growth process of CNTs even in other synthetic methods.

REFERENCES
[1]S. Iijima, *Nature*, **354**, 56 (1991).
[2]T. W. Ebbesen and P. M. Ajayan, *Nature*, **358**, 220 (1992).
[3]A. Thess, R. Lee, P. Nikolaev, H. Dai, P. Petit, J. Robert, C. Xu, Y.H. Lee, S.G. Kim, A.G. Rinzler, D.T. Colbert, G.E. Scuseria, D. Tomanek, J.E. Fischer, R.E. Smalley, *Science*, **273**, 483 (1996).
[4]T. Koyama, M. Endo, Y. Onuma, *Jpn. J. App. Phys.*, **11**, 445 (1972).
[5]M. Endo, K. Takeuchi, S. Igarashi, K. Kobori, M. Shiraishi, H. W. Kroto, *J. Phys. Chem. Solids*, **54**, 1841 (1993).
[6]P. Nilolaev, M. J. Bronikowski, R. K. Bradley, F. Rohmund, D. T. Colbert, K. A. Smith, R. E. Smalley, *Chem. Phys. Lett.*, **313**, 91 (1999).
[7]A. Maiti, C. J. Branbec, C. Ronald, J. Bernholc, *Phys. Rev.*, **B52**,14850 (1995).
[8]M. Yudasaka, T. Ichihashi, S. Iijima, *J. Phys. Chem.*, **B102**, 10201 (1998).
[9]H. Kataura, Y. Kumazawa, Y. Maniwa, Y. Ohtsuka, R. Sen, S. Suzuki, Y. Achiba, *Carbon*, **38**, 1691 (2000).
[10]N. Hamada, S. Sawada, A. Oshiyama, *Phys. Rev. Lett.*, **68**, 1579 (1992).
[11]R. Saito, M. Fujita, G. Dresselhaus, M.S. Dresselhaus, *Appl. Phys. Lett.*, **60**, 2204 (1992).
[12]J.W. Mintmire, B.I. Dunlap, C.T. White, *Phys. Rev. Lett.*, **68**, 631 (1992).
[13]T.W. Ebbesen, H.J. Lezec, H. Hiura, J.W. Bennett, H.F. Ghaemi, T. Thio, *Nature*, **382**, 54 (1996).
[14]J.W.G. Wildoer, L.C. Venema, A.G. Rinzler, R.E. Smalley, C. Dekker, *Nature*, **391**, 59 (1998).
[15]M. Kusunoki, M. Rokkaku, T. Suzuki, *Appl. Phys. Lett.*, **71**, 2620 (1997).
[16]M. Kusunoki, T. Suzuki, K. Kaneko, M. Ito, *Phil. Mag. Lett.*, **79**, 153 (1999).
[17]M. Kusunoki, T. Suzuki, T. Hirayama, N. Shibata, *Appl. Phys. Lett.*, **77**, 531 (2000).
[18]M. Kusunoki, T. Suzuki, C. Honjo, T. Hirayama and N. Shibata, *Chem. Phys. Lett.*, **366**, 458 (2002).

CARBURIZATION OF WC-CARBON NANOTUBE COMPOSITE USING C_2H_2 GAS

G. L. Tan,* X. J. Wu,[†] and Z. Q. Li[†]
*Department of Materials Science & Engineering, University of Pennsylvania
3231 Walnut St., Philadelphia, PA19104
[†]Department of Materials Science and Engineering, Zhejiang University
Hangzhou 310027, China

ABSTRACT
    An untraditional way to synthesize composites of nanophase WC powders and carbon nanotubes for the first time from the reduction and carburization of tungsten oxide nanoparticles by C_2H_2-H_2 gas instead of CO was investigated. Pure nanophase WC powders were fabricated through the carburization of pure WO_3 nano-powders by C_2H_2-H_2 gas with no Co in the precursors. Mixtures of α-W and WC were observed in the XRD patterns when the carburization temperature was below 750 °C, above which pure WC powders with a grain size of 10-30 nm were produced. When small amount of Co source was added into the tungsten precursors, carbon nanotubes and nanorods appeared in the composite powders after the same reduction and carbonization process. The effect of the process parameters, such as temperature and gas flow, on the final tungsten carbide phase has been studied in detail.

INTRODUCTION

    Hydrogen reduction of tungsten oxides is a complex, mixed diffusion-controlled reaction.[1, 2] During the reduction sequence, oxide intermediates are formed and transformed according to the temperature and dynamic oxygen partial pressure (humidity) within the powder layer. Generally, $WO_{2.9}$ and $WO_{2.0}$ are the intermediates of the dry H_2 reduction of WO_3 and will transform to α-W when temperature increases.[3] Tungsten carbide (WC) is traditionally produced by carburization of tungsten powder at high temperature with carbon black.[4] Carburization may be also accomplished using gaseous carbon monoxide.[5] Using solid carbon, it is easy to control the stoichiometry of the carbide powders at a high temperature (~ 1400 °C). In contrast, gaseous carburization agents provide the advantage of low carburization temperature.[6] Kear and McCandlish[7] created a new thermochemical processing, called thermal spray-fluid bed conversion process, to prepare WC-Co nanopowders as well as nanostructured WC-Co hard alloy. The microhardness was enhanced to 2260 N/mm^2 while the conventional value for the commercial WC-Co hard alloys in micrometer scale was within 1700~1900 N/mm^2. The carburization agent of this fluid bed conversion process was the mixture of CO/CO_2. Gao and Kear[8] reported a novel displacement reaction process to synthesize nanophase WC powder, which combines reduction and carburization of ammonium tungstate or tungsten oxide in a single operation at 700 °C for 2 hours in flowing H_2/CO. The particle size of the WC grains was estimated to be 10~20 nm. The same authors employed a TGA unit in situ monitoring and investigating the reduction and carburization process of tungsten oxide powders,[9] through which they found that reductive decomposition of WO_3 precursor powders resulted in high surface area α-W at temperatures above 650 °C, β-W at 575 °C. α-W was carburized at 700 °C to pure WC with average grain size of 9 nm, while β-W carburizing at 550 °C to WC with average grain size of 19 nm.

    Although CO is the most suitable agent for low temperature carburization and can avoid excess free carbon, it is more expensive than C_2H_2 gas. In order to wet the WC grains and sinter the tungsten carbide powders into compacted hard alloys, a small amount of Co powders need to be added into the WC matrix. Keep it in mind that when we prepare WC nanoparticles, Co can also be produced and mixed in the matrix in nanometer scale. Meanwhile, it is well known that at low temperature C_2H_2 could be decomposed into a lot of carbons, which could grow to carbon nanotubes *in situ* under the catalysis of Co nanoparticles. Therefore it was supposed that C_2H_2 gas could be used as a new carburiazation agent for preparation of composite of WC-Co nano-powders and carbon nanotubes *in situ* under the catalysis of cobalt nanoparticles. The propose was based on the idea that carbon nanotubes has exceptionally high Young's modules (more than 1.28 TPa), so they may be used to reinforce the matrix materials, i.e. improve the mechanical behavior of the matrix WC-Co alloys. In this paper, we would like to present for the first time how composite powders of WC nanoparticles and carbon nanotubes without detectable contaminant phases and free carbons have been successfully fabricated using C_2H_2 instead of CO as the carburization agent. Furthermore, we will continue on to investigate how these carbon nanotubes may improve the mechanical properties of sintered WC-Co hard alloys.

EXPERIMENTAL PROCEDURE

Ammonium metatungstate, $(NH_4)_6(H_2W_{12}O_{40})\cdot4H_2O$, was used as the tungsten precursor and $Co(NO_3)_2$ as the cobalt precursor. By drying ammonium metatungstate without Co salt, the precursor powders were calcined in a furnace at 300~450 °C in air for two hours. The precursor was decomposed into pure tungsten oxides nanopowders after calcination. Nanophase WO_3 precursor powders were also prepared from ammonium tungstate as described elsewhere.[8,9] About 10 g of nanometer WO_3 powders were heated in a quartz boat in a quartz tube furnace in a H_2 gas atmosphere (300 mL/min) with a heating rate of 5 °C/min. The WO_3 powders were reduced to α-W at a temperature range within 500-800 °C. After the reduction reaction, the atmosphere in tube system was immediately switched from the reduction gas (H_2) to the carburization agent (C_2H_2) and the furnace was heated up further to the carburization temperature above 800 °C. In this way, the high active α-W powders with large surface areas could be kept fresh without seeing air. The carburization process of α-W then started through the reaction of α-W with carbon decomposed from C_2H_2 gas at 700-900 °C for two hours. After carburization was accomplished, the tube system was gradually cooling at a rate of 2 °C/min to the room temperature.

The as-synthesized pure WC powders are hard to be densified into compacted alloys through hot-pressing sintering due to their high melting point. Therefore, it is necessary to introduce second phase to wet the grain during sintering. Cobalt and Nickel are most widely used as the wetting phase in hard alloy industry. We choose cobalt as the wetting phase for WC matrix. On the other hand, cobalt nanoparticles could catalyze the formation of carbon nanotubes from C_2H_2 gas decomposition in order to *in situ* form a composite of WC nanoparticles and carbon nanotubes. In this study, $Co(NO_3)_2$ with a weight percentage of 10 % ~ 20 % was mixed to ammonium metatungstate precursor and dried to paste powders. The composite WO_3 –CoO nanopowders were obtained after the paste precursor was calcined at 300~450 °C for one hour. Afterwards the mixture oxides powders were reduced to α-W (-Co) composite metal nanopowders by H_2 within the temperature of 500~800 °C. Finally α-W (-Co) composite nanopowders were carburized by C_2H_2 gas to WC-Co nanopowders. Under the catalysis of Co nanoparticles, the excess carbons being decomposed from C_2H_2 at high temperature grew up to carbon nanotubes or nanorods *in situ*. In this way, a composite of WC-Co nanoparticles and carbon nanotubes were fabricated simultaneously. The crystal structure development of the composite powers with carburization temperature was evaluated by X-ray diffraction (XRD) by taking samples after every step. The morphology and configuration of fabricated pure WC nanopowders or composite WC-Co-carbon nanotubes were investigated by transmission electron microscopy (TEM).

RESULTS AND DISCUSSION

Structural Development

To monitor the phase evolution during reduction and carburization, the process was stopped at various points and samples were taken out and stored in a glass vessel, prior to XRD and TEM measurements. The initial forms of the tungsten oxides after calcination of tungsten acid precursor below 550 °C in air were $WO_{2.0}$, $WO_{2.72}$ and $WO_{2.9}$. All of them had transferred to $WO_{2.9}$ and WO_3 when the calcination temperature increased to 600 °C for one hour.

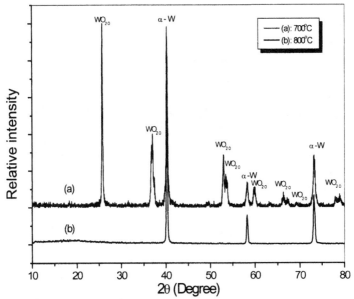

Figure 1. XRD patterns of the reduction products of nanophase WO_3 powders by H_2 within the temperature range of 700-800 °C for one hour. The reduction route of tungsten oxides is completed following path: $WO_3 \rightarrow WO_{2.9} \rightarrow WO_{2.72} \rightarrow WO_2 \rightarrow \alpha - W$.

The calcined tungsten oxide powders were then placed in a quartz boat, which was moved to the center of the tube furnace afterwards. The tube furnace was vacuumed, back fed with H_2 gas, and heated to a certain temperature for reduction. The reduced products were checked by XRD measurement, to determine at which temperature the tungsten oxides might be reduced to α-W. The results are shown in Figure 1. It can be seen that the product of the reduction process for WO_3 powders was pure α-W when temperature was above 800 °C, below which the mixture of $WO_{2.0}$ and α-W co-exist. The reaction duration at each temperature was set as one hour. However, even the reduction temperature was set as low as 500 °C for one hour, no β-W phase appeared in our products, which was different from Gao's result. The present reduction route was similar to that of Schubert's, which was completed along with the following path: $WO_3 \rightarrow WO_{2.9} \rightarrow WO_{2.72} \rightarrow WO_2 \rightarrow$ α-W when reduction temperature gradually increases. From Figure 1 and other XRD patterns for low temperature reduction products, it has been observed that our reduction route follows the same path: $WO_3 \rightarrow WO_{2.9} \rightarrow WO_{2.72} \rightarrow WO_2 \rightarrow \alpha - W$. The difference is that our reduction temperature is about 300 °C lower than that of Schubert's through the same reduction path. That's because WO_3 nanometer powders have much higher surface area and higher chemical activity than those conventional WO_3 coarse particles in micrometer scale. The higher the surface area, the more active the WO_3 nanopowders, and the lower the reduction temperature from active WO_3 to α-W by H_2. In addition, the diffuse path for H_2 into the particles is also much shorter for nanometer samples than that for micrometer particles. As a result, its reduction period was also shorter than that of micrometer powders. The mean grain size of α-W powders after reduction of tungsten oxides by H_2 at 800 °C for one hour was estimated to be 15 nm estimated from TEM images.

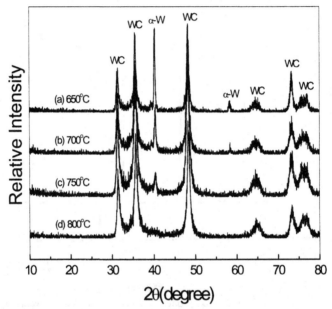

Figure 2. XRD patterns for the carburization products of α-W nanophase powders by C_2H_2/H_2 mixture gas instead of CO at different temperatures. Phase pure WC nano-powders were obtained by carburization of α-W at 800 °C.

After WO_3 nanophase powders were reduced to α-W by H_2, temperature of the quartz tube furnace increased up to the point at which the carburization of α-W occurred. In this case, a novel carburization process was chosen, in which the conventional carburization agent of CO gas was replaced with C_2H_2. The carburization temperature of C_2H_2 gas was higher than that of CO. There are lots of free carbon formed together with WC powders if the temperature and flow pressure of C_2H_2 gas were not carefully controlled.

The carburization of α-W by C_2H_2 will only be completed until the temperature reaches to 800 °C, below which α-W still remains in the carburization products. Figure 2 shows the XRD patterns for the carburization products of α-W by C_2H_2/H_2 mixtures instead of CO at different temperatures. It can be seen from Figure 2 that carburization of α-W by C_2H_2 was not completed within the temperature range of 650 °C to 750 °C. The lower the carburization temperature, the higher concentration the unreacted α-W phase and the less WC phase. Pure WC nanometer powders were obtained by C_2H_2 carburization of α-W at 800 °C as shown in Figure 2 (d), where the diffraction peaks from α-W disappeared. The average grain size of WC particles being carburized at 800 °C was estimated to be about 30 nm, which was estimated from TEM images (Figure 3). The particles agglomerated together forming strings of beads in nanometer scale. The light contrast of some particles is due to their deviation from focus plane, which indicates that those WC nanoparticles with light contrast are not located at the same height level with those with dark contrast.

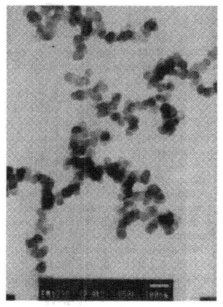

Figure 3. TEM image for pure WC nanoparticles (without Co), prepared by carburization of α-W nanoparticles through C_2H_2 agent at 800 °C for one hour.

Compared with Gao's results, it is found that the carburization temperature of α-W by C_2H_2 (800 °C) was about 100 °C higher than that by CO (700 °C), so the grain sizes also increase with the final carburization temperature. Therefore the particle size in our samples is bigger than that of Gao's sample, which was 16 nm. The reason that the carburization of α-W by C_2H_2 was higher than that by CO could be described as follows. In case of Kear's fluid bed conversion process using CO as carburization agent, the active tungsten sites react with the CO to form an incipient oxy-carbide carbide phase, liberating CO_2 gas that was also carried away by the flowing gas stream.[10] Further reaction gradually transforms the oxy-carbide into nanophase WC. There were no excess free carbons upon the surface of the particles during the carburization. In present study, using C_2H_2 gas as the carburization agent, the decomposed solid carbons can't be carried away by flowing gas stream. They stay inside the tube, mostly covering on the surface of the particles. The carburization process will continue even after the C_2H_2 is cutoff. Therefore the solid carbons in the inner carbon layer have to react with W to form WC on the surface of particle first, while the carbons on the outer layer will have to diffuse through the inner carbon layers and WC product layers to contact with the W atoms in the inner area inside the particles. The higher the reaction temperature, the faster and deeper the diffusion of carbon species, and the faster the carburization process. As a result, a high reaction temperature is necessary in order to complete the carburization. Therefore the mixture of α-W & WC would be the phase of final products at low carburization temperature as being shown in Figure 2 ((a)~(c)). Only when carburization temperature goes up to high enough (say 800 °C), then the diffusion coefficient may be large enough to drive the enough carbon atoms to pass through the product layers to reach the core area in order to make the W particle to be carburized completely. Figure 2 (d) demonstrate such a complete carburization results.

The additional XRD results for the carburization products under different gas flow of acetylene suggest that lower gas flow of acetylene produced metaphase W_2C due to the insufficient carbon source, while high gas flow of acetylene had induced the formation of WC and free carbon due to excess decomposed carbon. Only when the gas flow of acetylene was set to be suitable value (100 mL/min at 800 °C), the pure WC nanoparticles would be formed, as being demonstrated in Figure 2 (d).

It could also be seen from Figure 2 (d) that there exists no detectable contaminant phases, such as W_2C or free carbon, in the carburization product, suggesting that C_2H_2 could also be used as a suitable carburization agent for tungsten powders to produce tungsten carbides. This result exhibits an important prospect to replace expensive and toxic CO gas with cheap and safe C_2H_2 gas as carburization agent.

During the carburization process of the composite W- (Co) nanoparticles within the temperature of 650-800 °C, the carburization agent gas of C_2H_2 was decomposed into carbons which will be converted to carbon nanotubes or nanorods under the catalyst of free Co nanoparticles in the composite powders. In this way, the composite WC-Co nanoparticles and carbon nanotubes were *in situ* fabricated. Thus, the diffraction peak from free carbon was observed in the XRD patterns, which may contribute from the free carbon nanotubes as well as carbon nanorods in the composite nanoparticles.

Microstructure of WC-C Nanophase Powders

The microstructure of the composite nano-powders of WC-Co nanoparticles and carbon nanotubes is shown in Figure 4. Figure 4 (a) exhibits the mixture of aggregated WC nanoparticles with heavy contrast as well as some big carbon nanotubes, one of which is especially long and big with a diameter of about 30 nm and a length of more than 1 µm. Figure 4 (b) shows a cluster of carbon nanotubes inserting into the matrix of WC nanoparticles in heavy contrast, and the head of the cluster extends to outside the matrix powders. These WC nanoparticles are bounded with carbon nanotubes, possibly through chemical binding force. Therefore they are very useful to improve the mechanical properties of matrix WC hard alloys, because carbon nanotubes have exceptionally high Young's Modulus and strength.

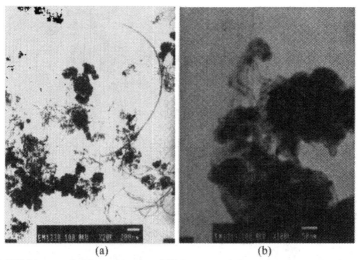

(a)	(b)

Figure 4. TEM images for the composite of WC nanoparticles and carbon nanotubes

Figure 5 (a) shows several big carbon "nano-bamboos" combined with other nanotubes and WC nanoparticles. Other small carbon nanotubes entangled together and then mixed with these "nano-bamboos" as well as WC nanoparticles. The diameter of these "nano-bamboos" is around 20 nm with pretty long length in c-direction, and several nano-bamboos shared two points. The section marks on these "nano-bamboos" are clearly seen, like those in real bamboos. Does these carbon "nano-bamboos" have high very toughness as real bamboos? If yes, they may be used to improve the toughness of the brittle materials. Figure 5(b) shows one big carbon nanorod in diameter of about 50 nm and more than 1 μm in length. Lots of WC nanoparticles aggregate around the nanorods forming branches of the tree. Because the contrast of the nanorod is much lighter than that of WC powders, therefore we may conclude that the rod is composed of carbon, instead of WC or Co heavy metals. Both the carbon nano-bamboos or coarse nanorods with exceptional high strength can be used to reinforce the matrix WC-Co hard alloys if they sintered together with these WC grains in wetting status.

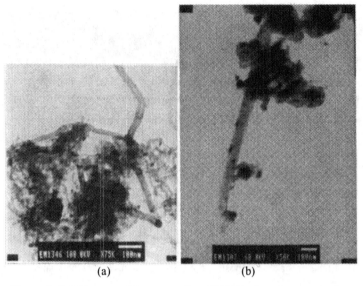

(a)	(b)

Figure 5. TEM images of the composite of WC-Co & carbon nanorods

SUMMARY

Pure phase WC nano-powders with grain sizes of 10-30 nm was successfully synthesized by choosing C_2H_2 instead of CO as the carburization agent, and no detectable second phase was observed in the XRD pattern. After introducing CO source in to tungsten precursor, the composite WC-Co nanopowders and carbon nanotubes or nanorods were obtained after carburization of the W–Co composite powders by C_2H_2 agent within the temperature range of 650-800 °C. The results show that the process parameters, such as carburization temperature and gas flow pressure, may have great effect on the purity of final products. By carefully controlling the gas flow, excess free carbon can be avoided and pure WC nanopowders were fabricated at 800 °C for two hours using C_2H_2 as carburization agent.

REFERENCES

[1] O. Kubaschewski, et al., *Metallurgical Thermaochemistry*, 5th ed., Pergamon Press (1979).

[2] D. R.Gaskill, *Introduction to Metallurgical Thermodynamics*, 2nd ed. McGraw-Hill (1981).

[3] W. D. Schubert, *Int. J. of Refeact, Metals & Hard Materials*, **9**, 178 (1990).

[4] Metals Handbook, 9th ed., Vol.7, ASM (1984).

[5] L. E. Toth, *Transition metal carbides and Nitrides*, Academic Press (1991).

[6] Y. T. Zhu and A. Manthiram, *J. Am. Ceram. Soc.*, **77**, 2777 (1994).

[7] B. H. Kear and L. E. McCandlish, *Nanostruct. Mater.*, **3**, 19 (1993).

[8] L. Gao and B. H. Kear, *NanoStructured Materials*, **9**, 205, 1997

[9] L. Gao and B. H. Kear, *Nanostruct. Mater.* **5**, 555(1995).

Environmental and Health Applications
and the Future of Nanotechnology

SYNTHESIS OF A BARIUM SULFATE NANOPARTICLE CONTRAST AGENT FOR MICRO-COMPUTED TOMOGRAPHY OF BONE MICROSTRUCTURE

Huijie Leng, Xiang Wang, Glen L. Niebur and Ryan K. Roeder
University of Notre Dame
Department of Aerospace and Mechanical Engineering
Notre Dame, IN 46556

ABSTRACT

A noninvasive, three-dimensional technique is needed to image microstructural features – such as microcracks and vasculature – in bone tissue. Micro-computed tomography (μCT) using a radiopaque barium sulfate contrast agent to label microstructural features in bone tissue has been proposed. The size of vasculature and microcracks in bone tissue requires nano-scale barium sulfate particles. Therefore, a simple aqueous precipitation method was used to synthesize barium sulfate particles less than 100 nm in size. The effects of the reagent solution concentration, molar ratio, feeding order, feeding speed and pH were investigated. Finally, the feasibility of imaging the synthesized barium sulfate nanoparticles in μCT was verified. The detected signal due to barium sulfate nanoparticles was amplified by a volumetric factor of nearly eight, which enabled detection of features much smaller than the nominal resolution of the instrument.

INTRODUCTION

Microdamage in bone tissue is caused by fatigue, creep, or monotonic overloading.[1] Accumulation of microdamage leads to a degradation of mechanical properties and an increased risk of fracture, including stress fractures in athletes and bone fragility in the elderly.[2-4] In human cortical bone, microdamage accumulates in the form of microcracks[5,6] and degradation of the elastic modulus is dependent on the applied stress state.[7,8] Therefore, quantitative measurement of the distribution and orientation of microcracks is important in order to assess the effects of microdamage on the mechanical behavior of bone tissue.

Microdamage in bone is currently imaged by optical microscopy using various epifluorescent contrast agents which chelate to calcium exposed at a free surface, such as a microcrack.[9-11] Sequential staining with multiple contrast agents can be used to differentiate microdamage from other microstructural features, as well as different damage events.[6,10,11] While bone tissue may be stained *en bloc*, quantitative measurements require the preparation of many histologic sections. Thus, these methods are inherently invasive and two-dimensional, and a non-invasive, three-dimensional technique is needed for imaging microdamage in bone. Micro-computed tomography (μCT) has been proposed for such a technique, using a suitable contrast agent.

In order for μCT to differentiate microcracks from bone, a contrast agent which is much more radiopaque than bone tissue must be used to selectively label microcracks. In practice, after staining with the contrast agent, microcracks would be detected as bright features in μCT images.

Barium sulfate ($BaSO_4$) is a logical choice for the contrast agent due to current clinical use as a contrast agent for conventional radiography of the digestive tract and a radiopaque filler in commercial bone cement. Thus, the biocompatibility of barium sulfate is reasonably well accepted. Typical microcracks in cortical bone tissue are 30-100 μm in length and less than one μm in width.[1,12,13] Therefore, in order to be used as a contrast agent for microcracks in cortical bone tissue, barium sulfate nanoparticles are required.

Barium sulfate precipitation has received significant attention in the literature, not only due to the industrial significance of scaling, but also as a model system for precipitation investigations.[15-24] Barium chloride and sodium sulfate are usually chosen as the reagents to study precipitation of barium sulfate. In some cases, sulfuric acid[15,21] or potassium sulfate[18-20] are also used. The size and morphology of any precipitated crystal is governed by the thermodynamics and kinetics of nucleation and growth.[25] Taguchi et al.[16] evaluated the nucleation and growth kinetics of barium sulfate for equimolar amounts of barium and sulfate ions, and Aoun et al.[18] studied the kinetics of barium sulfate precipitation with a molar excess of either barium or sulfate ions. Wong et al.[24] showed that a considerable excess of either reagent solution produced the smallest particles but the particles were all on the micro-scale.

Many methods and modifications have been used to control the size and morphology of precipitated barium sulfate, such as the addition of various additives,[21,26-28] chelate decomposition,[29] and microemulsions.[30-32] The use of additives, including chelating agents, has shown the greatest influence on particle morphology, but particle sizes have typically remained greater than nano-scale. Microemulsions also enable particle size control by varying the molar ratio of water to surfactant, but the uniformity of precipitates is more difficult to control and the yield is relatively small.

The objective of this work was to synthesize barium sulfate nanoparticles suitable for use as a contrast agent for μCT. Precipitation experiments focused on identifying key parameters for precipitating equiaxed nanoparticles. Finally, the feasibility of imaging the synthesized barium sulfate nanoparticles using μCT was investigated.

EXPERIMENTAL METHODS

Synthesis of Barium Sulfate Nanoparticles

Barium sulfate was precipitated instantaneously by mixing aqueous solutions of barium chloride and sodium sulfate as

$$BaCl_2 + Na_2SO_4 \rightarrow BaSO_4\downarrow + 2NaCl \qquad (1)$$

Barium chloride (Certified ACS crystal, Fisher Scientific, Fair Lawn, NJ) and sodium sulfate (anhydrous powder, Fisher Scientific, Fair Lawn, NJ) solutions were prepared by dissolving the desired concentration of reagents in de-ionized water. The pH of all reagent solutions was adjusted using nitric acid (ACS reagent, Aldrich Chemical Co., Milwaukee, WI) and sodium hydroxide (ACS reagent, Sigma Chemical Co., St. Louis, MO). Precipitation occurred as the feeding reagent solution was dropped into the other reagent solution using a buret to control the feeding rate. The mixed reagent solution was kept at ambient temperature and stirred at a constant speed of 450 rpm using a magnetic stir plate. Barium sulfate precipitates were collected by centrifugation at 4500 rpm for 20 min. Precipitates were washed with de-ionized water and recollected three times. Collected precipitates were dried in an oven at 40°C for more than 24 h.

Experiments were designed to study the effects of independent parameters on the size and morphology of precipitates. Experimental parameters included the reagent solution concentration, molar ratio, feeding order, feeding speed and pH. All experimental conditions are shown in Table I.

Powder x-ray diffraction (XRD) (X1 Advanced Diffraction System, Scintag, Inc., Sunnyvale, CA) was performed to verify the phase of the precipitated crystals. Powders were examined over 15-50° with a step size of 0.02° and a step time of 0.5 s, using Cu Kα radiation generated at 40 kV and 30 mA. The size and morphology of precipitates was examined qualitatively using field emission scanning electron microscopy (FESEM). For each sample, 0.003 g of precipitates was added to 1 ml methanol and dispersed in a sonic bath for 10 min. The suspension was dropped onto a sample holder and the methanol evaporated in air. After drying, samples were coated with gold and examined by FESEM (S-4500, Hitachi High-Technologies Corporation, Tokyo, Japan) with an accelerating voltage of 25 kV and working distance of 7 mm. The mean size of nanoparticles was also quantified by XRD line broadening measurements using the Scherrer equation and Warren's method.[33] The (200), (020) and (002) reflections were each scanned with a step size of 0.01° and a step time of 2.0 s, and fit with a Pearson 7 function using a freeware software package (MacDiff 4.2).

Table I. Experimental parameters investigated for barium sulfate precipitation.

Experiment	$[Ba^{2+}]$	$[SO_4^{2-}]$	$[Ba^{2+}]/[SO_4^{2-}]$	Feeding reagent	pH	Feeding speed
1	0.55 M	0.55 M	1.0	SO_4^{2-}	7	2.5 ml/min
2	0.01 M	0.01 M	1.0	SO_4^{2-}	7	2.5 ml/min
3	0.10 M	1.00 M	0.1	SO_4^{2-}	7	2.5 ml/min
4	1.00 M	0.10 M	10	SO_4^{2-}	7	2.5 ml/min
5	0.55 M	0.55 M	1.0	Ba^{2+}	7	2.5 ml/min
6	0.55 M	0.55 M	1.0	SO_4^{2-}	10	2.5 ml/min
7	0.55 M	0.55 M	1.0	SO_4^{2-}	7	25 ml/s

μCT of Barium Sulfate Nanoparticles

The barium sulfate nanoparticles synthesized using the parameters of Experiment 4 (Table I) were imaged using μCT. A cortical bone specimen was removed from the mid-diaphysis of a bovine tibia using a band saw. Six holes of 1.2 mm diameter and approximately 5.0 mm depth were drilled in the specimen. One hole was used as a control and the other five were filled with approximately 0.02 ml of aqueous suspensions containing 0.8, 0.6, 0.4, 0.2 and 0.1 vol% barium sulfate nanoparticles. After the water was absorbed by the bone, the specimen was dried and each of the six holes were imaged by μCT (80MG, Scanco Medical AG, Bassersdorf, Switzerland) at 10 μm resolution, 70 kVp voltage, 113 μA current intensity and 200 ms integration time. The image for each hole contained 141 slices (1.4 mm) where each slice was oriented parallel to the depth of the hole. A Gauss filter (sigma = 1; support = 2) was applied to reduce signal noise and a threshold brightness of 17,000 was selected based on the image data from the control hole. All voxels with brightness lower than the threshold were assumed to be cortical bone tissue and filtered off. All voxels with brightness greater than the threshold were assumed to be due to the barium sulfate nanoparticles. Three-dimensional images were constructed using the available software (Image Processing Language, Scanco Medical AG, Bassersdorf, Switzerland).

The minimum detectable thickness of deposited barium sulfate nanoparticles by μCT was also studied. Aqueous suspensions containing 0.2 and 0.1 vol% barium sulfate nanoparticles were prepared and 0.02 ml of each suspension was dropped onto a microscope slide. After the water evaporated, the diameter of the deposited layer was measured, and the thickness of the deposited layer was estimated assuming uniform deposition. Each glass slide was imaged by μCT using the methods described above, with slices normal to the deposited layer and a blank microscope slide as a control for threshold. The thickness of the deposited layer of barium sulfate nanoparticles on each slide was measured from μCT images after threshold.

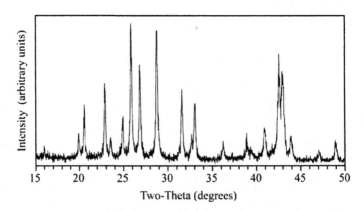

Fig. 1. X-ray diffraction pattern for barium sulfate nanoparticles synthesized with $[Ba^{2+}] = 1.0$ M, $[SO_4^{2-}] = 0.1$ M, SO_4^{2-} feeding, and feeding rate = 2.5 ml/min (Experiment 4).

RESULTS

Synthesis of Barium Sulfate Nanoparticles

XRD patterns identified precipitates from all experimental conditions as barium sulfate.[34] An exemplary XRD pattern for the barium sulfate nanoparticles prepared by Experiment 4 (Table 1) is shown in Fig. 1. The effects of the experimental parameters on the size and morphology of barium sulfate precipitates were revealed by FESEM. For equimolar amounts of each reagent, increased total reagent concentration resulted in decreased particle size and a more equiaxed morphology (Fig. 2a and b). Particles precipitated at 0.01 M reagent concentrations were plate-like and greater than 1 μm in size. At 0.55 M reagent concentrations, particles were equiaxed and submicron. For a fixed total reagent concentration, increasing the molar ratio of barium to sulfate ions resulted in a substantially decreased particle size (Fig. 2a, c and d). For a molar ratio of 10 (Experiment 4, Table 1), particles were less than 100 nm in diameter. For this powder, the mean particle size measured by XRD line broadening of the (200), (020) and (002) reflections was 52.0 nm, 54.5 nm and 46.3 nm, respectively. Thus, the particles were confirmed to be equiaxed with a mean diameter of approximately 50 nm. All other powders were measured to have mean diameters greater than 100 nm. The feeding reagent also affected the size and distribution of precipitates (Fig. 2a and e). For equimolar amounts of each reagent, particles prepared by sulfate feeding were significantly smaller and more uniform than those prepared by barium feeding. For equimolar amounts of each reagent and sulfate feeding, reduced feeding rate resulted in more uniform precipitates (Fig. 2a and f). The pH of reagent solutions was not shown to significantly affect the size or morphology of precipitates under the conditions investigated.

μCT of Barium Sulfate Nanoparticles

Varying amounts of barium sulfate nanoparticles were deposited in holes drilled in cortical bone tissue and imaged using μCT. Image segmentation at the threshold level of 17,000 removed the entire cortical bone image, and the higher intensity voxels representing barium sulfate were clearly visible (Fig. 3). Images revealed that the particles covered the inner wall of the holes such that the actual thickness of deposited barium sulfate nanoparticles was less than the hole diameter. The measured diameter of holes with deposited barium sulfate nanoparticles was larger than the actual diameter and the measured volume of deposited barium sulfate nanoparticles was larger than the actual volume of deposited particles. The measured volume of barium sulfate nanoparticles was nearly eight times larger than the actual volume (Fig. 4).

(a)   (b)

(c)   (d)

(e)   (f)

Fig. 2. SEM micrographs of the as-prepared barium sulfate precipitates showing the effects of selected experimental parameters: (a) Experiment 1, $[Ba^{2+}] = [SO_4^{2-}] = 0.55$ M, SO_4^{2-} feeding, feeding rate = 2.5 ml/min; (b) Experiment 2, $[Ba^{2+}] = [SO_4^{2-}] = 0.01$ M, SO_4^{2-} feeding, feeding rate = 2.5 ml/min; (c) Experiment 3, $[Ba^{2+}] = 0.10$ M, $[SO_4^{2-}] = 1.0$ M, SO_4^{2-} feeding, feeding rate = 2.5 ml/min; (d) Experiment 4, $[Ba^{2+}] = 1.0$ M, $[SO_4^{2-}] = 0.1$ M, SO_4^{2-} feeding, feeding rate = 2.5 ml/min; (e) Experiment 5, $[Ba^{2+}] = [SO_4^{2-}] = 0.55$ M, Ba^{2+} feeding, feeding rate = 2.5 ml/min; (f) Experiment 7, $[Ba^{2+}] = [SO_4^{2-}] = 0.55$ M, SO_4^{2-} feeding, feeding rate = 25 ml/s.

Three glass slides with different amounts of deposited barium sulfate nanoparticles were also imaged using μCT. For the 0.2 vol% barium sulfate suspension, the imaged thickness of the barium sulfate layer was 30 μm and the diameter of the layer was about 8 mm, which was almost the same as the measured diameter. For the 0.1 vol% barium sulfate suspension, no layer of pixels was detected. The estimated layer thickness for the 0.2 and 0.1 vol% barium sulfate suspensions was 0.79 μm and 0.24 μm, respectively.

(a)   (b)

Fig. 3. Three-dimensional μCT images of barium sulfate nanoparticles deposited from an (a) 0.8 vol% and (b) 0.1 vol% suspension into 1.2 mm diameter holes in cortical bone tissue.

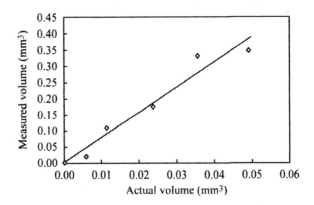

Fig. 4. The relationship between the actual volume of deposited barium sulfate nanoparticles and the volume measured by μCT after threshold. Linear regression of the data yielded $y = 7.75x$ ($R^2 = 0.95$).

DISCUSSION

Synthesis of Barium Sulfate Nanoparticles

The kinetics of precipitation under stoichiometric conditions is well understood. Increasing nucleation rate or decreasing crystal growth rate generally results in smaller precipitates. In this investigation barium sulfate precipitated instantaneously upon mixing the two reagent solutions and the entire system was kept under constant stirring. Therefore, nucleation was expected to govern the precipitate size more than crystal growth. The nucleation rate is influenced by the supersaturation,[18] and the supersaturation ratio, S, is defined as

$$S = \sqrt{\left[Ba^{2+}\right]\left[SO_4^{2-}\right]/K_s} \qquad (2)$$

where the solubility product of barium sulfate is $K_s \approx 1.1 \cdot 10^{-10}$.[24] Thus, increasing total reagent concentration increases the supersaturation and nucleation rate. A higher nucleation rate means that a greater number of stable nuclei form before reaching equilibrium, limiting crystal growth and resulting in a smaller mean particle size.

Accordingly, the results of this study showed that the size of barium sulfate precipitates was reduced from micro-scale to nano-scale by increasing the total reagent concentrations (Fig. 2a and b). A further decrease in particle size was achieved using a molar excess of barium to sulfate (Fig. 2a and d). This result can also be explained by considering the local supersaturation as the sulfate feeding reagent solution was added dropwise to the barium reagent solution. In the case for a molar excess of barium, a higher supersaturation was maintained throughout the duration of precipitation. In other words, the local supersaturation incipient to precipitation was lower for the equimolar reagent solutions, and was governed by the larger volume of the barium reagent solution rather than the droplets of the feeding reagent solution. Furthermore, as barium sulfate precipitated upon mixing the equimolar reagent solutions, the supersaturation decreased rapidly. This decrease in local supersaturation with precipitation was even greater for a molar shortage of barium to sulfate (molar ratio of 0.1). Thus, decreasing molar ratio resulted in an increased variation in the supersaturation during precipitation and a greater variation in the precipitate size distribution (Fig. 2a, c and d). Other investigations have also reported this effect, including a more pronounced effect for a molar excess of barium versus a molar excess of sulfate due to preferential ion adsorption.[18, 24] In order to investigate the latter effect in our system, further experiments are needed where the feeding reagent is reversed for the same molar ratios. Furthermore, the particle size distributions should be quantified in future work. Nonetheless, for equimolar concentrations of barium and sulfate, sulfate feeding resulted in somewhat smaller, more uniform precipitates than barium feeding (Fig. 2a and e), suggesting that a molar excess of barium increased the nucleation rate more than an equal molar excess of sulfate. The increased precipitate size distribution observed for a faster feeding rate (Fig. 2a and f) can be explained by less uniform mixing (greater variation in local supersaturation) for the faster feeding rate. Finally, under the conditions investigated, the solution pH had negligible effect on the supersaturation and therefore showed no effect on the particle size.

μCT of Barium Sulfate Nanoparticles

The results of this study showed the feasibility of detecting barium sulfate nanoparticles in cortical bone tissue using μCT. Furthermore, the results also indicated that barium sulfate nanoparticles influenced adjacent image voxels. The detected signal was amplified by a volumetric factor of nearly eight (Fig. 4). In other words, barium sulfate nanoparticles enabled detection of features smaller than the nominal resolution of the instrument. Currently most commercially available μCT scanners have a resolution on the order of 10 μm and clinical CT scanners have a resolution on the order of 100 μm. Microcracks in bone are typically 30-100 μm in length and no more than 1 μm in width.[1, 12, 13]. Therefore, if barium sulfate nanoparticles can be delivered to microcracks, the amount of barium sulfate nanoparticles or the size of microcracks may have dimensions less than the resolution of the instrument. Moreover, the results from imaging barium sulfate nanoparticles deposited on glass slides indicated a minimum detectable dimension between 0.24 μm and 0.79 μm for barium sulfate nanoparticles using an instrument with 10 μm resolution.

ACKNOWLEDGEMENTS

This work was supported by the National Institutes of Health grant AR 49598.

REFERENCES

[1]R.B. Martin, "Fatigue microdamage as an essential element of bone mechanics and biology," *Calcif. Tissue Int.*, **73** [2] 101-107 (2003).

[2]M. B. Schaffler, E. L. Radin and D. B. Burr, "Mechanical and morphological effects of strain rate on fatigue of compact bone," *Bone*, **10** [3] 207-214 (1989).

[3]M.R. Forwood and A.W. Parker, "Microdamage in response to repetitive torsional loading in the rat tibia," *Calcif. Tissue Int.*, **45**, 47-53 (1989).

[4]D. B. Burr, M.R. Forwood, D.P. Fyhrie, R.B. Martin, M.B. Schaffler and C.H. Turner, "Bone microdamage and skeletal fragility in osteoporotic and stress fractures," *J. Bone Miner. Res.*, **12**, 6-15 (1997).

[5]P. Zioupos and A. Casinos, "Cumulative damage and the response of human bone in two-step loading fatigue," *J. Biomechanics*, **31**, 825-833 (1998).

[6]F.J. O'Brien, D. Taylor and T.C. Lee, "Microcrack accumulation at different intervals during fatigue testing of compact bone," *J. Biomechanics*, **36**, 973-980 (2003).

[7]K.J. Jepsen, D.T. Davy and D.J. Krzypow, "The role of the lamellar interface during torsional yielding of human cortical bone," *J. Biomechanics*, **32**, 303-310 (1999).

[8]D. Vashishth, K.E. Tanner and W. Bonfield, "Fatigue of cortical bone under combined axial-torsional loading," *J. Orthop. Res.*, **19**, 414-420 (2001).

[9]D. B. Burr and M. Hooser, "Alterations to the *en bloc* basic fuchsin staining protocol for the demonstration of microdamage produced *in vivo*," *Bone*, **17**, 431-433 (1995).

[10]T.C. Lee, E.R. Myers and W.C. Hayes, "Fluorescence-aided detection of microdamage in compact bone," *J. Anat.*, **193**, 179-184 (1998).

[11]T.C. Lee, T.L. Arthur, L.J. Gibson and W.C. Hayes, "Sequential labelling of microdamage in

bone using chelating agents," *J. Orthop. Res.*, **18**, 322-325 (2000).

[12]D.B. Burr, C.H. Turner, P. Naick, M.R. Forwood, W. Ambrosius, H. M. Sayeed and R. Pidaparti, "Does microdamage accumulation affect the mechanical properties of bone?," *Journal of Biomechanics*, **31**, 337-345 (1998).

[13]D. Taylor and T.C. Lee, "Measuring the shape and size of microcracks in bone," *J. Biomechanics*, **31**, 1177-1180 (1998).

[14]D.J. Gunn and M.S. Murthy, "Kinetics and mechanisms of precipitation," *Chem. Eng. Sci.*, **27**, 1293-1212 (1972).

[15]M.S. Murthy, "Theory of crystal growth in phase transformations: precipitation of barium sulfate," *Chem. Eng. Sci.*, **49**, 2389-2393 (1994).

[16]K. Taguchi, J. Garside and N.S. Tavare, "Nucleation and growth kinetics of barium sulphate in batch precipitation," *J. Crystal Growth*, **163**, 318-328 (1996).

[17]M.A. vanDrunen, H.G. Merkus and G.M. vanRosmalen, "Barium sulfate precipitation: Crystallization kinetics and the role of the additive PMA-PVS," *Part. Part. Syst. Char.*, **13** [5] 313-321 (1996).

[18]M. Aoun, E. Plasari, R. David and J. Villermaux, "Are barium sulphate kinetics sufficiently known for testing precipitation reactor models?," *Chem. Eng. Sci.*, **51**, 2449-2458 (1996).

[19]M. Aoun, E. Plasari, R. David and J. Villermaux, "A simultaneous determination of nucleation and growth rates from batch spontaneous precipitation," *Chem. Eng. Sci.*, **54**, 1161-1180 (1999).

[20]B. Bernard-Michel, M.N. Pons and H. Vivier, "Quantification, by image analysis, of effect of operational conditions on size and shape of precipitated barium sulphate," *Chem. Eng. J.*, **87**, 135-147 (2002).

[21]M. Yokota, E. Oikawa, J. Yamanaka, A. Sato and N. Kubota, "Formation and structure of round-shaped crystals of barium sulfate," *Chem. Eng. Sci.*, **55**, 4397-4382 (2000).

[22]R. Aguiar, H. Muhr and E. Plasari, "Comparative study of the influence of homogeneous and heterogeneous (multi-phase) precipitation processes on the particle size distribution," *Chem. Eng. Technol.*, **26** [3] 292-295 (2003).

[23]L. Vicum, M Mazzotti and J. Baldyga, "Applying a thermodynamic model to the non-stoichiometric precipitation of barium sulfate," *Chem. Eng. Technol.*, **26**, 325-333 (2003).

[24]D.C.Y. Wong, Z. Jaworski and A.W. Nienow, "Effect of ion excess on particle size and morphology during barium sulphate precipitation: an experimental study," *Chem. Eng. Sci.*, **56**, 727-734 (2001).

[25]A.G. Walton, *The Formation and Properties of Precipitates*, Robert E. Krueger Publishing Co., Huntington, NY, 1979.

[26]S.N. Black, L.A. Bromley, D. Cottier, R.J. Davey, B. Dobbs and J.E. Rout, "Interactions at the organic/inorganic interface: Binding motifs for phosphonates at the surface of barite crystals," *J. Chem. Soc. Faraday Trans.*, **87**, 3409-3414 (1991).

[27]F. Jones, A. Stanley, A.L. Rohl and M.M Reyhani, "The role of phosphonate speciation on the inhibition of barium sulfate precipitation," *J. Crystal Growth*, **29**, 584-593 (2003).

[28]L.A. Bromley, D. Cottier, R.J. Davey, B. Dobbs and S. Smith, "Interactions ate the

organic/inorganic interface: Molecular design of crystallization inhibitors for barite," *Langmuir*, **9**, 3594-3599 (1993).

[29]M. Uchida, T. Sue, A. Yoshioka and A. Okuwaki, "Hydrothermal synthesis of needle-like barium sulfate using a barium(II)-EDTA chelate precursor and sulfate ions," *J. Mater. Sci. Lett.*, **19**, 1373-1374 (2000).

[30]L. Qi, J. Ma, H. Cheng and Z. Zhao, "Preparation of $BaSO_4$ nanoparticles in non-ionic w/o microemulsions," *Colloid Surface A*, **108** 117-126 (1996).

[31]M. Li and S. Mann, "Emergence of morphological complexity in $BaSO_4$ fibers synthesized in AOT microemulsions," *Langmuir*, **16**, 7088-7094 (2000).

[32]J.D. Hopwood and S. Mann, "Synthesis of barium sulfate nanoparticles and nanofilaments in reverse micelles and microemulsions," *Chem. Mater.*, **9**, 1819-1828 (1997).

[33]B.D. Cullity, *Elements of X-Ray Diffraction*, 2nd Edition, Addison-Wesley Publishing Co., Inc., Reading, MA, 1978.

[34]Powder Diffraction File 5-0448, JCPDS-International Center for Diffraction Data (ICDD), 1993.

INCREASED SURFACE AREA AND ROUGHNESS PROMOTES OSTEOBLAST ADHESION ON HYDROXYAPATITE/TITANIA/PLGA COMPOSITE COATINGS

Michiko Sato and Elliott B. Slamovich
School of Materials Engineering
Purdue University
501 Northwestern Avenue
West Lafayette, IN 47907

Thomas J. Webster
School of Materials Engineering and
Department of Biomedical Engineering
Purdue University
1296 Potter Building
West Lafayette, IN 47907

ABSTRACT

The objective of this study was to elucidate mechanisms of increased osteoblast adhesion on hydroxyapatite (HA; $Ca_{10}(PO_4)_6OH_2$) composite coatings. Previous results showed increased osteoblast adhesion on HA coatings after hydrothermal treatment. In these studies, titanium was coated with HA, TiO_2, and Poly (dl-lactic-glycolic acid) (PLGA) using sol-gel processing. The HA particles synthesized were composed of agglomerated nano-sized crystallites. Surface features of the HA/TiO_2/PLGA composite coating possessed sizes in the micrometer and nanometer regime. The present study revealed increased surface area/roughness for the coatings after hydrothermal treatment, which may be a contributing factor to the previously observed enhanced osteoblast adhesion.

INTRODUCTION

Calcium phosphorous compounds like HA have been used for orthopedic/dental implant applications. HA-coated implants may promote sufficient bone formation directly on juxtaposed bone and thus establish a firm fixation between the bone and the implant.[1] Moreover, TiO_2 also has excellent biocompatibility. For example, Ramires et al. demonstrated that a composite of HA/TiO_2 stimulated osteoblast (bone forming cells) function more than uncoated titanium.[2] Since osteoblast adhesion is a necessary requirement for subsequent calcium deposition and increased bonding of an implant to juxtaposed bone, HA coatings exhibiting properties which stimulate osteoblast adhesion is desired. For instance, surface topography in both the nanometer and micrometer regime has been shown to improve osteoblast adhesion.[3,4]

Many different techniques including plasma spray processing, sol-gel processing, and electrochemical deposition have been proposed to coat titanium with HA. Techniques are desired to produce HA coatings which have few cracks and exceptional properties. Although plasma spray deposition is a widely used technique, it is performed at high temperatures resulting in multiple phases and lower crystallinity HA.[5] It has been reported that the dissolution rate of HA increases with decreasing crystallinity,[6] and that other calcium phosphorous compounds dissolve faster than HA.[7] Compared to plasma spray, sol-gel processing can attain a more uniform composition and surface morphology with fewer cracks. Sol-gel processing is also suitable to coat complex shapes.[8] In our previous studies,[9] the HA/TiO_2/PLGA composite coating demonstrated more osteoblast adhesion than a plasma-sprayed HA coating. The surface chemistry of the coating may have influenced osteoblast adhesion. Specifically, the presence of HA and TiO_2 stimulated osteoblast adhesion. After hydrothermal treatment in water or HA-precipitate aqueous solution at 70°C, osteoblast adhesion increased when compared to respective coating formulations before hydrothermal treatment. However, in addition to changes in surface chemistry, changes in surface area may have also influenced osteoblast adhesion. For this reason, the objective of this study was to correlate changes in surface area/roughness of the HA/TiO₂/PLGA composites to alterations in previously observed osteoblast adhesion.

EXPERIMETAL PROCEDURES

For preparation of the novel HA-based coatings, ammonium phosphate (Sigma) was added to deionized water (dH_2O) to produce a 0.60 M solution and the solution pH was adjusted to 10 using ammonium hydroxide (Fisher Scientific, Inc.). A 1.0 M calcium nitrate solution (Sigma) was then added dropwise to the ammonium phosphate solution at a rate of 18 ml/min with stirring. Precipitates appeared as soon as the calcium nitrate solution was added. The solution was stirred for 24h at room temperature. The precipitates were analyzed with XRD (Siemens D500 Kristalloflex; Bruker AXS Inc.), a light-scattering particle-size analyzer (COULTER LS 230; Coulter Corporation), and a BET surface area analyzer (SA3100; Beckman Coulter). HA crystallite size was calculated using: $D=6/(\rho S)$, where ρ is density and S is specific surface area as determined by BET. A HA theoretical density of 3.156 g/cm^3 was used. After precipitation, the HA powders were rinsed with dH_2O. Ethanol and chloroform were then added respectively to rinse the HA powders. Chloroform was then added to the HA powders; this solution was labeled as A (Figure 1). Poly (dl-lactic-glycolic acid) (PLGA; 50:50 wt% PLA:PGA; molecular weight = 12,000-16,500; Polysciences) and a TiO₂ precursor (titanium diisopropoxide bis(acetylacetonate); Aldrich) were dissolved in chloroform; this solution was labeled as B. PLGA was chosen since it will degrade as new bone grows, thus a locking mechanism between the coating and juxtaposed bone may increase with time. Solution A was then mixed with solution B. The ratio of each component was PLGA: TiO₂: HA = 3.5:4.6:2.0 by weight, assuming complete hydrolysis.

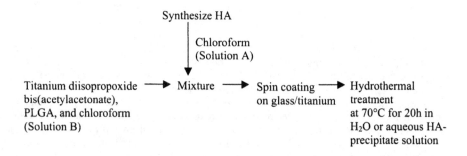

Figure 1. Flow sheet for sol-gel coating preparation

Glass slides (Becton Dickinson and Company) were used as substrates (1 cm × 1 cm × 0.1 cm). The glass substrates were ultrasonically washed with acetone and ethanol, respectively, and then rinsed with dH_2O. The mixture of solution A and B was deposited onto glass using a spin coater (EC101DT; Headway Research, Inc.). Coatings were dried in air at room temperature. Some of the coatings were then treated hydrothermally in dH_2O at 70°C for 20h, and dried at room temperature. The HA sol-gel coating formulation (produced according to the methods described previously) was also deposited on titanium (1 cm × 1 cm × 0.2 cm, Alfa Aesar Co.). The titanium pieces were sandblasted with 36 – 60 mesh aluminum oxide before coating. Some of the HA sol-gel coatings on titanium were treated hydrothermally in the aqueous solution containing the HA powders obtained earlier. This was done to prevent Ca and P from leaching out during hydrothermal treatment. Coatings were characterized using atomic force microscopy (AFM, Multimode SPM; Digital Instruments), to analyze surface morphologies, root mean square (RMS) roughness, and surface area. The AFM tip (NP-20; Veeco Instruments) had a nominal tip radius of curvature of 20-60 nm. Scanning was conducted on five different places chosen randomly at a scanning rate of 1 Hz in a contact mode. Obtained surface areas and RMS values were averaged and reported with standard deviation (SD).

RESULTS AND DISCUSSION

From Figure 2(A), precipitates dried in room temperature before hydrothermal treatment exhibited broad peaks. After hydrothermal treatment at 70°C for 20h, the peaks were shaper indicating a greater degree of crystallinity and did not show other phases except for HA (Figure 2(B)).

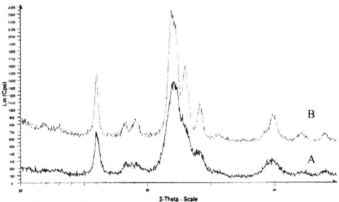

Figure 2. XRD of HA containing solution stirred for 24h: A) after dried at room temperature and B) after hydrothermal treatment at 70°C for 20h and dried at room temperature.

Figure 2. XRD of HA containing solution stirred for 24h: A) after dried at room temperature and B) after hydrothermal treatment at 70°C for 20h and dried at room temperature.

The mean particle size was 5.8 ± 1.8 (standard deviation) μm for the non-dried precipitates before hydrothermal treatment. The crystallite size for HA particles before and after hydrothermal treatment at 70°C for 20h was 15 nm and 18 nm, respectively. This indicates that the nano-sized HA was highly agglomerated when it was mixed with PLGA and the TiO_2 precursor.

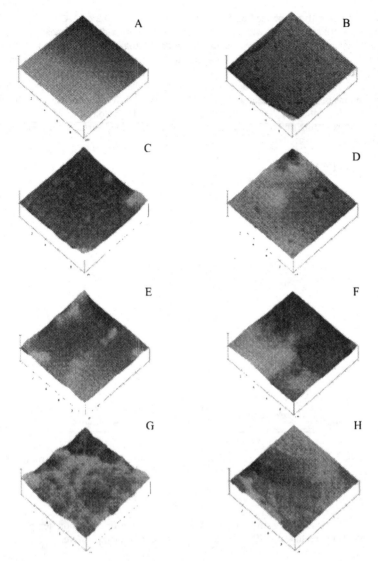

Figure 3. AFM scans of glass coated with: (A) PLGA before hydrothermal treatment and (B) after hydrothermal treatment; (C) PLGA + TiO$_2$ precursor before hydrothermal treatment and (D) after hydrothermal treatment; (E) PLGA + TiO$_2$ precursor + HA before hydrothermal treatment and (F) after hydrothermal treatment; titanium coated with: (G) PLGA + TiO$_2$ precursor + HA before hydrothermal treatment and (H) after hydrothermal treatment. Projected area is 10 × 10 μm.

Table I. Surface roughness of substrates and coatings

Substrates or coatings	RMS ± SD (nm)	Surface area ± SD (μm^2) (Projected area is 100 μm^2)	% of surface area compared to base substrate
Glass	18 ±12	102 ± 1.2	-
PLGA on glass before hydrothermal treatment	15 ± 7	100 ± 0.0	98*
PLGA on glass after hydrothermal treatment	116 ± 91	112 ± 5.8	110*
PLGA + TiO$_2$ precursor on glass before hydrothermal treatment	272 ± 69	114 ± 8.4	112*
PLGA + TiO$_2$ precursor on glass after hydrothermal treatment	124 ± 30	140 ± 21.0	137*
PLGA + TiO$_2$ precursor + HA on glass before hydrothermal treatment	205 ± 77	118 ± 7.5	116*
PLGA + TiO$_2$ precursor + HA on glass after hydrothermal treatment	213 ± 68	115 ± 8.7	113*
Titanium	134 ± 56	108 ± 2.7	-
PLGA + TiO$_2$ precursor + HA on titanium before hydrothermal treatment	308 ± 53	141 ± 11.8	131**
PLGA + TiO$_2$ precursor + HA on titanium after hydrothermal treatment	301 ± 109	179 ± 17.5	166**

* compared to glass; ** compared to titanium

The surface morphology of the coatings was characterized using AFM. The PLGA coating before hydrothermal treatment had the least RMS roughness and lowest surface area of all coatings (Figure 3(A)). The PLGA coating after hydrothermal treatment increased the surface area by 12 % compared to the respective coating formulation before hydrothermal treatment (Table I). This indicates that the surface of PLGA was decomposed due to hydrolysis and eroded (Figure 3(B)). The PLGA + TiO$_2$ precursor coating before hydrothermal treatment appeared to have bumps and some micro cracks, while the PLGA + TiO$_2$ precursor coating after hydrothermal treatment possessed pores (Figure 3(C) and (D), respectively). Compared to the respective coating before hydrothermal treatment, the surface area increased by 23 %. The surface area increased because of the creation of pores and cracks after hydrothermal treatment and maybe a relatively small contribution from peptization of a TiO$_2$ precursor as a result of hydrolysis. There were micro-sized islands on the surface of the PLGA + TiO$_2$ precursor + HA coatings before and after hydrothermal treatment (Figure 3(E) and (F)). This is consistent with the findings that the mean size of the HA particles was on the micrometer scale after precipitation. HA and TiO$_2$ particles were exposed on the surface of the PLGA + TiO$_2$ precursor

+ HA coating after hydrothermal treatment while the surface of particles were coated with PLGA for the respective formulation before hydrothermal treatment. This hypothesis is supported by the decomposition of the PLGA coating after hydrothermal treatment and because smaller islands/grains became more visible on TiO_2 and $HA/TiO_2/PLGA$ coatings after hydrothermal treatment. PLGA + TiO_2 precursor + HA coating on titanium after hydrothermal treatment exhibited the largest surface area and RMS roughness since titanium substrates exhibited a higher amplitude of roughness than glass substrates. Moreover, differences of surface area/roughness of the PLGA + TiO_2 precursor + HA coating on glass and titanium substrates could be attributed to the wettability of each substrate, thus, the coating formulation may have spread out better on glass than titanium. In addition, the surface area increased after hydrothermal treatment of the PLGA + TiO_2 precursor + HA coating on titanium by 27% compared to the respective coating before hydrothermal treatment. It is not clear why the same formulation of coating on glass did not have increased surface area after hydrothermal treatment.

Results of osteoblast adhesion tests from our previous studies[9] and present studies indicate that osteoblast adhesion increased with increasing surface area of coatings when comparing the same coating formulation before and after hydrothermal treatment. However, it is important to note that the extent that physical features and surface chemistry cooperate to promote osteoblast adhesion has not yet been elucidated. The influence of surface chemistry on osteoblast adhesion is confirmed by comparing the PLGA + TiO_2 precursor + HA on glass before and after hydrothermal treatment because the amplitude of roughness and surface area for these two coatings were similar. The PLGA + TiO_2 precursor coated glass after hydrothermal treatment exhibited a statistically significant greater surface area than the PLGA + TiO_2 precursor + HA coated glass after hydrothermal treatment although a statistically significant difference in osteoblast adhesion density was not observed between these two substrates.

CONCLUSIONS

The $HA/TiO_2/PLGA$ composite coating using sol-gel technique produced a non-cracked coating and previously demonstrated more osteoblast adhesion when compared to an HA plasma-sprayed coating. Results from this study showed that synthesized HA particles were composed of agglomerated crystallites of 15nm and 18nm before and after hydrothermal treatment, respectively. Therefore, the HA/TiO_2 composite in PLGA coating possessed surface roughness both in the micron and nanometer regime. Hydrothermal treatment even at low temperatures promoted crystallization of HA and increased surface area on all coatings except for the PLGA + TiO_2 + HA on glass. PLGA + TiO_2 and PLGA + TiO_2 + HA coatings exposed smaller grains/islands on the surface after hydrothermal treatment when compared to the respective coating before hydrothermal treatment due to the decomposition of PLGA. Since previous studies demonstrated that hydrothermal treatment increased osteoblast adhesion on coatings, this study suggests that increased surface area of the hydrothermally-treated coatings may be a reason why. Such information is crucial to developing the next generation of orthopedic implant coatings.

ACKNOWLEDGEMENTS

The authors would like to thank the Whitaker Foundation for financial assistance and Elizabeth Massa-Schlueter for research assistance.

REFERENCES

[1]K. Søballe, S. Overgaard, E.S. Hansen, M.Lind, C. Bunger, "A Review of Ceramic Coatings for Implant Fixation," *J. Long-Term Eff. Med. Implt.*, **9** [1822] 131 (1999).

[2]P.A. Ramires, A. Romito, F. Cosentino, E. Milella, "The Influence of Titania/Hydroxyapatite Composite Coatings on In Vitro Osteoblast Behaviour," *Biomaterials*, **22**, 1467-74 (2001).

[3]T.J. Webster, C. Ergun, R.H. Doremus, R.W. Siegel, R. Bizios, "Enhanced Functions of Osteoblasts on Nanophase Ceramics," *Biomaterials*, **21**, 1803-10 (2000).

[4]K. Anselme, M. Bigerelle, B. Noël, A. Iost, P. Hardouin, "Effect of Grooved Titanium Substratum on Human Osteoblastic Cell Growth", *J. Biomed. Mater. Res.*, **60**, 529-40 (2002).

[5]L. Sun, C.C. Berndt, K.A. Gross, A. Kucuk, "Materials Fundamentals and Clinical Performance of Plasma-Sprayed Hydroxyapatite Coatings: A Review," *J. Biomed. Mater. Res.*, **58**, 570-92 (2001).

[6]Y.L. Chang, D. Lew, J.B. Park, J.C. Keller, "Biomechanical and Morphometric Analysis of Hydroxyapatite-Coated Implants With Varying Crystallinity," *J. Oral. Max. Surg.*, **57** [9] 1096-1108 (1999).

[7]L. Cleries, J.M. Fernandez-Pradas, G. Sardin, J.L. Morenza, "Dissolution Behaviour of Calcium Phosphate Coatings Obtained by Laser Ablation," *Biomaterials*, **19** 1483-87 (1998)

[8]D.M. Liu, T. Troczynski, D. Hakimi, "Effect of Hydrolysis on the Phase Evolution of Water-Based Sol-Gel Hydroxyapatite and Its Application to Bioactive Coatings," *J. Mater. Sci.: Mater. Med.*, **13**, 657-65 (2002).

[9]M. Sato, E.B. Slamovich, T.J. Webster, "Enhanced Osteoblast Adhesion on a Novel Hydroxyapatite Coating," *Mat. Res. Soc. Symp. Proc.*, **774**, O7.33.1-6 (2003).

IMPROVED BONE CELL ADHESION ON ULTRAFINE GRAINED TITANIUM AND Ti-6Al-4V

Chang Yao and Elliott B. Slamovich
School of Materials Engineering
Purdue University
West Lafayette, IN 47907

Javaid I. Qazi and Henry. J. Rack
School of Materials Science and Engineering
Clemson University
Clemson, SC 29634

Thomas J. Webster
Department of Biomedical Engineering and School of Materials Engineering
Purdue University
West Lafayette, IN 47907

ABSTRACT

Nanotechnology involves the use of materials whose components exhibit novel and significantly changed properties when control is gained at the atomic, molecular, and supramolecular level. Although nanomaterials have been revolutionizing traditional science and engineering disciplines (e.g., catalytic, mechanical, electrical), advantages of nanophase materials in biological applications remain to date largely uninvestigated. The objective of this *in vitro* study has been to determine osteoblast (bone-forming cell) functionality on ultrafine grained metal surfaces. Results provide evidence of increased osteoblast adhesion on ultrafine commercial purity titanium and Ti-6Al-4V compared to respective conventional grain size metals after 4 hour exposure. Since adhesion is a prerequisite for deposition of calcium-containing bone-forming minerals, further enhancement is expected on ultrafine grained metals. This study therefore adds ultrafine/nanophase metals to the growing list of materials wherein enhanced bone cell function pertinent to successful orthopedic implant applications has been observed.

INTRODUCTION

Metallic materials have been extensively used for biomedical device applications. For example, stainless steels, cobalt-chromium alloys, as well as titanium and its alloys have been used for orthopedic implants. Among them, titanium and its alloys, because of their spontaneous formation of a highly biocompatible titanium dioxide passive film following exposure to air and/or blood, their excellent corrosion resistance against both atmospheric and aggressive fluidic environments, their high strength to weight ratio and their relatively low modulus of elasticity when compared to other metallic implants, are increasingly utilized for biodevice fabrication.[1] Currently, three titanium alloy sub-groups, commercial purity titanium, $\alpha + \beta$ titanium alloys, and metastable β titanium, are used for bone plates, screws, total hip and knee replacements, etc.[2]

Clearly, the implant surface, because it is in direct contact with living tissue, plays an important role in any interaction between an implant and its surroundings.[3,4] Several studies have suggested that four materials-related factors can influence an implant's biocompatibility and ultimately integration into juxtaposed bone. These include surface chemical composition, surface energy, surface wettability, and topography.[5-8] Since bone is composed of constituent nano-scale components, e.g., collagen and hydroxyapatite, it might be imagined that implants with nano or near nanophase surface features may enhance new bone growth.

It is further thought that the scale of a material's grain size, surface roughness and topography should influence the functionality of cells on an implant surface. For example, previous studies have found increased osteoblast (bone-forming cell) functionality on nanophase ceramics (alumina, titania, hydroxyapatite), polymers, carbon fibers, and composites compared to their respective conventional counterparts.[9-13] In contrast, similar investigations of nanophase metals on osteoblast behavior are still in their infancy.[14,15] Indeed, only one study has been reported cataloguing the increased osteoblast functionality on nanophase vis a conventional commercial purity titanium, Ti-6Al-4V, and CoCrMo where the nanophase material was prepared by cold compaction of nanograined powders.[14] While this study suggests the benefits that might be expected from nanograined metal implants, the rather low mechanical integrity of the materials examined is not expected to be of interest to clinicians. The present study has therefore examined osteoblast adhesion on ultrafine/nanophase commercial purity titanium and Ti-6Al-4V prepared by severe plastic deformation. Indeed, this process has shown itself to be capable of producing ultrafine grained commercial purity titanium and Ti-6Al-4V with materially enhanced mechanical performance when compared to conventional wrought alloys.[16]

MATERIALS AND METHODS
Substrates

Cell adhesion studies were performed on four types of substrates: commercial purity titanium, Ti-6Al-4V, titanium foil and borosilicate glass. The first two categories provided a direct comparison between ultrafine grained and conventional grain size materials, while the borosilicate glass, which was etched in 1N NaOH for 1 hour prior to inclusion in the culture environment, served as a reference substrate. The ultrafine grained samples had been prepared by combined equal channel angular extrusion and cold drawing/extrusion while the conventional biomedical grade alloys were provided by ALLVAC, Monroe, NC. In each instance they were machined to 1 cm diameter, 2.5 mm thick discs and mechanically polished through 1000 grit SiC paper. In contrast, the titanium foil (99.2% pure from Alfa Aesar Inc.) was in annealed condition with conventional grain size, this being cut into 1 cm x 1 cm squares for testing.

All substrates were soaked in acetone and 70 % ethanol, ultrasonically cleaned, dried in an oven and finally sterilized in an autoclave for 30 minutes just prior to commencement of the cell culture studies.

Material Characterization

Phase analysis of the titanium alloys was carried out by x-ray diffraction analysis using CuKα radiation (Scintag XDS 2000 θ-θ diffractometer). Similarly, microstructure analysis was performed using scanning and scanning transmission electron microscopy (Hitachi 3500 SEM; Hitachi HD 2000 STEM). Samples for SEM analysis were prepared by cold mounting followed by standard grinding and polishing. Microstructure was revealed by etching the sample in a 2 ml H_2O_2 + 2 ml HF + 95 ml H_2O solution. SEM was operated in secondary electron imaging (SEI) mode to analyze the microstructure. Thin foils for STEM were prepared by cutting and grinding slices ~150μm thickness. 3 mm discs were then punched out from these slices and dimpled from both sides to obtain ~25 μm thickness in the center. Finally, thin areas were obtained by electropolishing these dimpled foils in a solution consisting of 300 ml methanol, 175 ml butanol and 35 ml perchloric acid at -50 to -55°C under a voltage of 24V.

Cell Culture and Adhesion Test

Human osteoblasts (CRL-11372 American Type Culture Collection) were used for the cell adhesion experiments at population numbers between 9~11. Substrates were placed in 12-well cell culture clusters and rinsed with phosphate buffered saline (PBS). The cells were cultured in Dulbecco's Modified Eagle Medium (DMEM) supplemented with 10% fetal bovine serum (Hyclone) and 1% Penicillin/Streptomycin (Hyclone) and were seeded at a density of 3500 cells/cm². Cells were then allowed to adhere under standard cell culture conditions (a humidified, 5% CO_2 / 95% air environment at 37°C).

Following exposure for 4 hours, the culture medium was removed from all wells and the substrates were gently rinsed with PBS three times to remove any non-adherent cells. The adherent cells were then fixed with 4% formaldehyde solution and stained with Hoescht dye (Sigma). The cell numbers in the field of view (a circle with 2.16 mm diameter under 100X magnification) were counted under a fluorescence illumination (Leica), the amount of fluorescence detected approximately correlating with the number of adherent cells on the substrate surface. Five random fields were counted per substrate with all experiments run in triplicate and repeated three separate times; data were analyzed using standard analysis of variance (ANOVA) procedures followed by a Student's T-test. Statistical significance was considered at $p < 0.1$.

RESULTS

As can be seen in Fig.1, SEM micrographs showed that the titanium foil exhibited a grain size of approximately 10 μm while the biomedical grade titanium and Ti-6Al-4V exhibited grain sizes of about 20μm and 5μm, respectively. In contrast, STEM pictures showed that the ultrafine/nano grained commercial purity titanium and Ti-6Al-4V exhibited grain sizes between 100-200 nm.

XRD results (as seen in Fig.2) confirmed that the commercial purity titanium, independent of grain size, was single phase alpha having an h.c.p. crystal structure, while Ti-6Al-4V was a two-phase alpha plus beta (b.c.c.) alloy. Further examination of the x-ray spectra showed a difference in crystallographic texture between the conventional and ultrafine grained materials, this difference being attributed to the processing routes utilized to prepare these materials. Finally, line broadening was noted in the ultrafine grained materials, the degree of line broadening being characteristic of fine grained alloys. Finally, XRD analysis also showed that the titanium foil consists of single hcp (α) phase.

Fig.3 demonstrates that osteoblast adhesion increased 76% and 73% for ultrafine grained commercial purity titanium and Ti-6Al-4V, respectively, compared to titanium foil. At the same time, cell densities for conventional commercial purity titanium and Ti-6Al-4V were only about 15% greater than that of the titanium foil. More importantly, osteoblast adhesion increased significantly ($p < 0.1$) on the ultrafine grained metals compared to titanium foil while osteoblast adhesion was similar ($p > 0.1$) when comparing the conventional metals with the titanium foil. As a result, it can be concluded that osteoblast adhesion increased on the ultrafine grained metals when compared to conventional titanium alloys. It is important to note that the grain sizes in ultrafine grained commercial purity titanium / Ti-6Al-4V (100~200nm) were much smaller than that of either the titanium foil (10μm) or conventional titanium (20μm) or Ti-6Al-4V (~5μm) alloys.

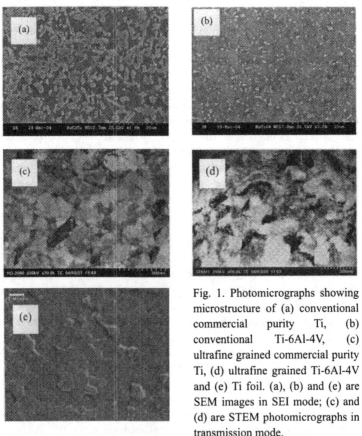

Fig. 1. Photomicrographs showing microstructure of (a) conventional commercial purity Ti, (b) conventional Ti-6Al-4V, (c) ultrafine grained commercial purity Ti, (d) ultrafine grained Ti-6Al-4V and (e) Ti foil. (a), (b) and (e) are SEM images in SEI mode; (c) and (d) are STEM photomicrographs in transmission mode.

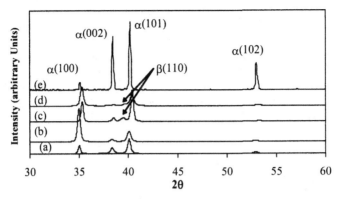

Fig. 2. X-ray diffraction patterns for (a) conventional commercial purity titanium, (b) ultrafine grained commercial purity titanium, (c) conventional Ti-6Al-4V, (d) ultrafine grained Ti-6Al-4V and (e) titanium foil.

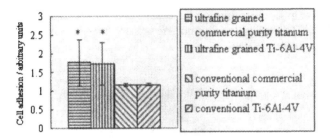

Fig. 3. Enhanced osteoblast adhesion on ultrafine grained titanium compared to conventional commercial purity titanium and Ti-6Al-4V, cell adhesion results being normalized to adhesion on wrought Ti foil. Data are mean $\pm$ SEM; n = 3; $*p < 0.1$ compared to the titanium foil.

DISCUSSION

The underlying mechanism of increased cell adhesion on ultrafine grained metals is still under investigation. However, several unique characteristics of these materials may be contributing to increased osteoblast adhesion. For example, the ratio of grain boundary to matrix volume increases with decreasing grain size. Additionally, the grain boundary misorientation is a function of the thermomechanical history of the ultrafine grained materials.[16] Collectively, these features may lead to either an increased number of reactive sites or an increase in reactive site activity thereby promoting cell adhesion. Moreover, cell adhesion to biomaterials is primarily influenced by the interactions between surface bound proteins and corresponding receptors on the membrane of the cells.[17,18] For this reason, specific types of grain boundaries in ultrafine grained materials may also serve as reactive sites promoting those protein interactions that are important for osteoblast adhesion. Indeed this suggestion is supported by previous studies of nanophase metals prepared by powder-metallurgy techniques, which have shown osteoblast adherence is enhanced at grain boundaries.[14]

It's important to note that future *in vitro* studies should also be needed to look at cell differentiation, protein synthesis and other phenomena. However, since we know that the adhesion of osteoblasts on an implant surface is a prerequisite for subsequent deposition of calcium-containing minerals found in bone, these results suggest that ultrafine grained titanium and its alloys may be attractive materials for use in orthopedic applications.

CONCLUSION

The present study provides evidence of increased osteoblast adhesion on ultrafine grained/nanophase commercial purity titanium and Ti-6Al-4V when compared to their conventional grain size counterparts. One possible explanation is that surface structure of an ultrafine grained/nanograined material contains an ever increasing proportion of disturbed grain boundary volume with decreasing grain size. It is hypothesized that the titanium oxide film resident above these grain boundary regions is different from the titanium oxide normally present above the "matrix" – this surface disturbance may promote protein interactions and subsequent cell adhesion. Since adhesion is a prerequisite for subsequent osteoblast functionality, these results suggest that enhanced osteointegration could be possible through substitution of ultrafine/nanophase metals in orthopedic applications.

ACKNOWLEDGEMENTS

The authors would like to thank the State of Indiana 21st Century Funds for financial support.

REFERENCES
[1]D.M. Brunette, P. Tengvall, M. Textor and P. Thomsen, Foreword in *Titanium in Medicine: Material Science, Surface Science, Engineering, Biological Responses and Medical Applications*, 1st ed. Springer, Germany, 2001.
[2]G. Lütjering and J.C. Williams, "Biomedical Applications," pp.345-49 in *Titanium*, 1st ed. Edited by B. Derby. Springer, Germany, 2001.
[3]M. Jayaraman, U. Meyer, M. Bühner, U. Joos and H.P. Wiesmann, "Influence of Titanium Surfaces on Attachment of Osteoblast-like Cells In Vitro," *Biomaterials*, 25 [4] 625-631 (2004).
[4]R. Lange, F. Lüthen, U. Beck, J. Rychly, A. Baumann and B. Nebe, "Cell-extracellular Matrix Interaction and Physico-chemical Characteristics of Titanium Surfaces Depend on the Roughness of the Material," *Biomolecular Engineering*, 19 [2-6] 255-261 (2002).
[5]Z. Schwartz and B.D. Boyan, "Underlying Mechanisms at the Bone-biomaterial Interface," *J. Cell Biochem.*, 56, 340-347 (1994).
[6]X. Zhu, J. Chen, L. Scheideler, R. Reichl and J. Geis-Gerstorfer, "Effects of Topography and Composition of Titanium Surface Oxides on Osteoblast Responses," *Biomaterials*, In Press, Available online.

[7]H. Huang, "In Situ Surface Electrochemical Characterizations of Ti and Ti–6Al–4V Alloy Cultured with Osteoblast-like Cells," *Biochemical and Biophysical Research Communications,* **314** [3] 787-792 (2004).

[8]P.T. de Oliveira and A. Nanci, "Nanotexturing of Titanium-based Surfaced Upregulates Expression of Bone Sialoprotein and Osteopontin by Cultured Osteogenic Cells," *Biomaterials,* **25**, 403-413 (2004).

[9]T.J. Webster, "Nanophase Ceramics as Improved Bone Tissue Engineering Materials," *American Ceramic Society Bulletin,* 82 [6].

[10]M. Karlsson, E. Pålsgård, P.R. Wilshaw and L. Di Silvio, "Initial In Vitro Interaction of Osteoblasts with Nano-porous Alumina," *Biomaterials,* **24** [18] 3039-3046 (2003).

[11]T.J. Webster, R.W. Siegel and R. Bizios, "Osteoblast Adhesion on Nanophase Ceramics," *Biomaterials,* **20** [13] 1221-1227 (1999).

[12]N.R. Washburn, K.M. Yamada, C.G. Simon, S.B. Kennedy and E.J. Amis, "High-throughput Investigation of Osteoblast Response to Polymer Crystallinity: Influence of Nanometer-scale Roughness on Proliferation," *Biomaterials,* **25** [7-8] 1215-1224 (2004).

[13]R.L. Price, L.G. Gutwein, L. Kaledin, F. Tepper and T.J. Webster, "Osteoblast Function on Nanophase Alumina Materials: Influence of Chemistry, Phase, and Topography," *Journal of Biomedical Materials Research,* 2004.

[14]T.J. Webster and J.U. Ejiofor, "Increased Osteoblast Adhesion on Nanophase Metals," *Biomaterials,* in press, available on-line.

[15]T.J. Webster and J.U. Ejiofor, "Increased, Directed Osteoblast Adhesion at Nanophase Ti and Ti and Ti6Al4V Particle Boundaries," *Materials Research Society Symposium Proceedings* MM10.4 (2003).

[16]R.Z. Valiev, V.V. Stolyarov, H.J. Rack and T.C. Lowe, "SPD-processed Ultra-fine Grained Ti Materials for Medical Applications," *pp. 362-367, Medical Device Materials, ASM,* Materials Park, OH, 2004.

[17] K. Anselme, "Osteoblast Adhesion on Biomaterials", *Biomaterials,* 21 [7] 667-681 (2000).

[18]Y. Yang, R. Glover and J.L. Ong, "Fibronectin Adsorption on Titanium Surfaces and Its Effect on Osteoblast Precursor Cell Attachment," *Colloids and Surfaces B: Biointerfaces,* **30** [4] 291-297 (2003).

IMPROVED DISPERSION OF NANOPHASE TITANIA IN PLGA ENHANCES OSTEOBLAST ADHESION

Huinan Liu and Elliott B. Slamovich
School of Materials Engineering
Purdue University
West Lafayette, IN 47907

Thomas J. Webster
Department of Biomedical Engineering and School of Materials Engineering
Purdue University
West Lafayette, IN 47907

ABSTRACT
Much work is needed in the design of more effective bone tissue engineering materials since the average lifetime of an orthopedic implant is less than 15 years. Frequently, orthopedic implants fail due to insufficient integration into juxtaposed bone. Nanotechnology offers exciting alternatives to traditional bone implants since bone itself is a nanostructured material composed of nanofibered hydroxyapatite well-dispersed in a mostly collagen matrix. For this reason, osteoblast (bone-forming cell) adhesion on poly-lactic-co-glycolic acid (PLGA) and nanophase titania composites was investigated *in vitro*. For this purpose, PLGA was dissolved in chloroform and 30 wt. % nanometer grain size titania was dispersed by various sonication powers from 0 W to 332.5 W. The surface characteristics of the composites were studied by scanning electron microscopy. Results showed that the dispersion of titania in PLGA was significantly enhanced by increasing the intensity of sonication. Moreover, for the first time, results correlated greater osteoblast adhesion with increased nanophase titania dispersion in PLGA. Such promising results may be due to greater percentages of titania at the surfaces of composites prepared under higher sonication. In this manner, the present study demonstrates that PLGA composites with well-dispersed nanophase titania can improve osteoblast functions necessary for increased orthopedic implant efficacy.

INTRODUCTION
The scientific challenge of bone regeneration encompasses not only understanding cell functions but also the development of suitable scaffold materials that act as templates for cell adhesion, growth and proliferation. Several physicochemical and biological requirements have to be fulfilled by the scaffold, depending on the particular application under consideration. Specifically, for orthopedic applications, scaffolds should have the following characteristics: (i) biocompatible and bioresorbable with a controllable degradation and resorption rate to match cell/tissue growth *in vitro* and/or *in vivo*; (ii) suitable surface chemistry and roughness for cell attachment, proliferation and differentiation; and (iii) bioactivity and osteoconductivity to facilitate the migration of osteoblasts from surrounding bone into the implant site and hence assist in the healing process.[1-3] To satisfy these demanding criteria, investigators have been studying a wide variety of natural and synthetic biomaterials, like polymers and ceramics, for design and construction of scaffolds for orthopedic tissue engineering. These include naturally occurring polymers (e.g. hydrogels like gelatin, fibrin or collagen[4-6]), synthetic bioresorbable

polymers (e.g. polylactic acid, polyglycolic acid and poly-lactic-co-glycolic acid[7-9]), bioactive ceramics (e.g. bioglasses and hydroxyapatite derivatives[10-12]) and naturally occurring ceramics (such as coral[13]). As a means to repair defects in bone, the design of polymer/ceramic composites offers an exceptional opportunity to combine biodegradability and bioactivity to optimize scaffolds for bone regeneration with tailored physical and biological properties. The addition of a ceramic phase to a biodegradable polymer may be exploited to alter the polymer degradation behavior towards favorable directions.[14] Moreover, functions of bone-forming cells can be increased in polymer/ceramic composites from better cell seeding and growth environments due to improved osteoconductivity provided by the bioactive ceramic phase.[15-20]

In attempts to further emulate the composite nature of bone, a number of groups have been investigating the use of nanophase ceramics (or ceramics with grain sizes less than 100 nm). Specifically, reports in the literature demonstrated enhanced functions (adhesion, proliferation, synthesis of alkaline phosphatase, deposition of calcium-containing mineral, etc.) of osteoblasts on nanophase ceramics that mimic the grain size of physiological bone, i.e. less than 100 nm in diameter. For example, alumina, titania and hydroxyapatite have been tested in this manner.[21-23] Bone itself is a nanostructured material composed of materials like proteins (collagen Type I) and hydroxyapatite crystals that have nanometer dimensions. For example, the dimensions of crystalline hydroxyapatite in natural bone are from 50-100 nm in length and 1-10 nm in diameter. Thus, it stands to reason that osteoblasts are naturally accustomed to interacting with nanostructured surface roughness in the body. Therefore, one promising consideration for the next generation of orthopedic implants with improved efficacy is degradable polymer and nanophase ceramic composites.

For these reasons, the objective of the present *in vitro* study was to investigate osteoblast adhesion on polymer/nanophase ceramic composites. Poly-lactic-co-glycolic acid (PLGA) was chosen as the model polymer in the current study since it is biodegradable and widely utilized in tissue engineering applications. Nanophase titania was utilized as the model ceramic due to its excellent biocompatibility.[21-22] Since nanophase ceramics tend to significantly agglomerate when added to polymers, different sonication output powers were investigated in this study to enhance ceramic dispersion. In addition, the functions of osteoblasts were determined on PLGA/titania composites formulated at different sonication powers. Osteoblasts are primarily responsible for synthesizing mineralized bone matrix. Osteoblasts adhere on the surface of bone, proliferate and secrete extracellular matrix proteins, as well as deposit calcium mineral.

MATERIALS AND METHODS
Substrate Preparation

Polymer/ceramic composites: PLGA pellets (50/50 wt. % polylactic/glycolic acid; Polysciences, Warrington, PA) were dissolved in 8 mL of chloroform at 50 °C in a water bath for 40 minutes. Nanophase titania powder (Nanophase Technologies, Romeoville, IL) was then added to the PLGA solution to give a 70/30 polymer/ceramic weight ratio. The purity of the titania powder was 99.5+%, the particle size was 32 nm according to BET measurements and the crystal phase was 80% anatase/20% rutile. The composite suspension was then either used without sonication, or sonicated using a W-380 sonicator (Heat System – Ultrasonics, Inc.) with output power settings of 118.75 W to 332.5 W. After sonication, the suspension was cast into a Teflon petri dish, evaporated in air at room temperature for 24 hours and dried in an air vacuum chamber (50.8 kPa) at room temperature for 48 hours. Finally, the composite films (0.2 mm in thickness) were cut into 1×1 cm squares for cell experiments.

Polymer films: PLGA pellets were dissolved in 8 mL chloroform, cast into a Teflon petri dish, evaporated in air at room temperature for 24 hours and vacuum dried for 48 hours at room temperature. Then, the PLGA films (0.18 mm in thickness) were cut into 1×1 cm squares and used as polymer control substrates.

Ceramic compacts: Green titania (TiO_2) circular disks were prepared by dry pressing nanophase titania powders in a tool-steel die (10 mm in diameter) via a uniaxial pressing cycle from 0.6 to 3 GPa over a 10 min period into 0.5 mm in thickness. Then some disks were sintered by heating in air at the rate of 10 °C/min from room temperature to a final temperature of 600 °C, holding at 600 °C for 2 hours and cooling down at the same rate as the heating rate. Both green and sintered titania disks were used as ceramic control substrates.

Reference materials: Glass coverslips were etched in 1 N NaOH and prepared for experiments according to standard protocols. Glass was used as a reference material.[24] All material samples were degreased in acetone and ethanol according to established laboratory procedures.

Sterilizing procedures: Composite and PLGA substrates were sterilized by soaking in ethanol for 20 minutes before performing experiments with cells. Titania substrates were sterilized by exposing them to UV light for 1 hour for each side.

Table I. PLGA, TiO_2 and PLGA/TiO_2 composites substrates

Substrate	Parameters
PPC	Pure PLGA, control
TCG	Green pure titania compacts, control
TCS	Sintered pure titania compacts, control
PTC0	PLGA/titania composites without sonication
PTCL	PLGA/titania composites sonicated at low power for 10 min*
PTC25	PLGA/titania composites sonicated at 118.75 W for 10 min
PTC35	PLGA/titania composites sonicated at 166.25 W for 10 min
PTC45	PLGA/titania composites sonicated at 213.75 W for 10 min
PTC70	PLGA/titania composites sonicated at 332.5 W for 10 min

* Composites sonicated using a standard biological sonicator (VWR Aquasonic Cleaner Model 50-T)

Surface Characterization

Surface topographies of the PLGA/titania composites were characterized according to standard scanning electron microscopy (SEM) techniques using a JEOL JSM-840 Scanning Electron Microscope at a 5 kV accelerating voltage. Substrates were first sputter-coated with a thin layer of gold-palladium using a Hummer I Sputter Coater (Technics) in a 100 millitorr vacuum in an argon environment for three minutes with a 10 mA current. SEM images taken at 10 kX and 40 kX magnifications were used to determine topographical differences between the substrates of interest to the present study.

Cytocompatibility Testing
Cell culture: Human osteoblasts (bone-forming cells; CRL-11372 American Type Culture Collection) were cultured in Dulbecco's modified Eagle's medium (DMEM; GIBCO, Grand Island, NY) supplemented with 10% fetal bovine serum (FBS; Hyclone) and 1% penicillin/streptomycin (P/S; Hyclone) under standard cell culture conditions, that is, a sterile, 37 °C, humidified, 5% CO_2/95% air environment. Cells were used without further characterization.

Cell adhesion: Osteoblasts were seeded at a density of 2500 cells/cm^2 onto the substrates and were incubated under standard cell culture conditions for 4 hours. Cells were then fixed with formaldehyde (Fisher Scientific, Pittsburgh, PA) and stained with Heoscht 33258 dye (Sigma); the cell nuclei were visualized and counted under a fluorescence microscope. Cell counts were expressed as the average number of cells on eight random fields of substrate surface area. For all experiments, etched glass coverslips were used as reference substrates. All experiments were run in triplicate and repeated at least three separate times.

Cell morphology: Osteoblast morphology on the PLGA/titania composites was examined using an environmental scanning electron microscopy (ESEM) at about 5 torr (≈ 660 Pa) in a hydrated state at 10 keV.

Statistical analysis: Cell densities were analyzed using standard analysis of variance (ANOVA) techniques; statistical significance were considered at $p < 0.05$.

RESULTS
Substrate Topography
    Results of the present study provided evidence of increased nanometer surface roughness and subsequent surface area of PLGA/TiO_2 composites fabricated with higher sonication power. Titania particles with different sizes were visible on the top and bottom sides of the composites, respectively, as shown in figure 1. Scanning electron micrographs suggest that the distribution of ceramic particles was different on the top and bottom side of composite films depending on sonication power utilized, and there was more agglomeration on the bottom than on the topside when comparing the composites made under the same sonication power. Larger ceramic agglomerations tended to settle down in the polymer solution first and, thus, composites had larger amounts of ceramic at the bottom at lower sonication power.

    The amount of surface area occupied by titania increased on the topside of the composite films, while it decreased on the bottom side of the films when the sonication intensity was increased. Specifically, 10% compared to 5.7% of the surface area occupied was titania on the topside at 166.25 W and 118.75 W sonication power, respectively. Also, large agglomerations of nanophase titania particles were broken, and titania particles were more evenly dispersed in the PLGA matrix with higher sonication powers.

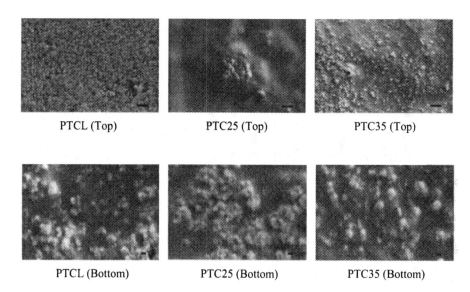

| PTCL (Top) | PTC25 (Top) | PTC35 (Top) |
| PTCL (Bottom) | PTC25 (Bottom) | PTC35 (Bottom) |

Figure 1. SEM micrographs of PLGA/titania thin films with low intensity sonication (PTCL), PLGA/titania thin films with 25% maximum output sonication power (118.75 W) (PTC25), and PLGA/titania thin films with 35% maximum output sonication power (166.25 W) (PTC35) on the top side and bottom side of the films. Original magnification: 10 kX for top side and 40 kX for bottom side; magnification bars (lower right): 1 μm.

Cell Adhesion

Due to the differences in topography between the top and bottom sides of the composites, osteoblast adhesion tests were conducted on both sides. Osteoblast adhesion on the top side of PLGA/titania composites, the pure PLGA films (PPC), the pure green titania compacts (TCG), the pure sintered titania compacts (TCS) and the glass reference are shown in figure 2. Osteoblast adhesion was significantly ($p < 0.01$) greater on the topside of the PTC35 substrate than the PTC0, PTCL, and PTC25 substrates after 4 hours. PTC35 was the PLGA/titania composite made with 35% maximum sonication power, which equaled 166.25 W. Osteoblast adhesion on the pure green titania compacts was significantly ($p < 0.05$) greater than most other substrates, including glass, PPC, PTC0, PTCL, PTC25, PTC45, and PTC70, but interestingly, not significantly greater than PTC35.

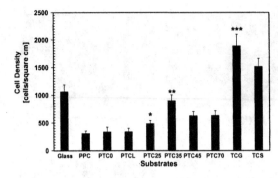

Figure 2. Osteoblast adhesion on the topside of PLGA/titania composite films. 1) Glass, 2) PPC, 3) PTC0, 4) PTCL, 5) PTC25, 6) PTC35, 7) PTC45, 8) PTC70, 9) TCG, and 10) TCS. Values are mean ± SEM; n = 3; *p < 0.05 compared to PTC0 and PTCL; **p < 0.01 compared to PTC0, PTCL, and PTC25; ***p < 0.05 compared to PPC, PTC0, PTCL, PTC25, PTC45, and PTC70.

Osteoblast adhesion on the bottom side of PLGA/titania composites is shown in figure 3. Osteoblast adhesion on the bottom side of the composites decreased with increasing sonication power since more ceramic particles distributed to the topside. Osteoblast adhesion was significantly (p < 0.01) smaller on the bottom side of PTC35 substrate than the PTC0, PTCL, and PTC25 after 4 hours.

Comparing the top with bottom sides of the composites, at lower sonication power, osteoblast adhesion was better on the bottom side; while at higher sonication power, osteoblast adhesion was better on the topside of the composites. When sonication power was greater than 166.25 W, osteoblast adhesion was significantly (p < 0.01) greater on the topside of the composites than the respective bottom side.

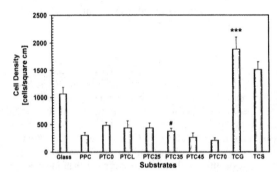

Figure 3. Osteoblast adhesion on the bottom side of PLGA/titania composite films. 1) Glass, 2) PPC, 3) PTC0, 4) PTCL, 5) PTC25, 6) PTC35, 7) PTC45, 8) PTC70, 9) TCG, and 10) TCS. Values are mean ± SEM; n = 3; #p < 0.05 compared to PTC45 and PTC70; ***p < 0.05 compared to PPC, PTC0, PTCL, PTC25, PTC45, and PTC70.

Cell Morphology on the Substrates

Micrographs of osteoblast adhesion on the substrates were obtained using an ElectroScan ESEM 2020, as shown in figure 4. The cells seeded on both green (TCG) and sintered titania (TCS) compacts were more well-spread than on the surface of the PLGA/titania composite (PTC35) and pure PLGA. In addition, the cells seeded on the surface of the composite were better spread than on the pure PLGA. Since increased cell spreading indicates a more favorable surface for osteoblast interactions, these results confirm those presented in figures 2 and 3.

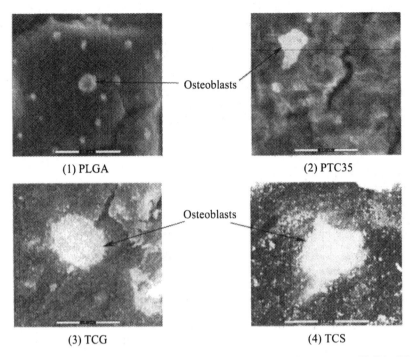

Figure 4. ESEM micrographs of osteoblast adhesion on the (1) pure PLGA (PPC), (2) PLGA/titania composite film (PTC35), (3) green titania compact (TCG), and (4) sintered titania compacts (TCS) substrates after 4 hours of incubation. Magnification bars: 10 μm.

DISCUSSION

Results of this study demonstrated that the dispersion of nanophase titania in PLGA was significantly enhanced by increasing the intensity of sonication. That is, higher ultrasonic energy broke large titania agglomerations into smaller titania particles, which were more easily dispersed in PLGA suspensions.

Moreover, results correlated greater osteoblast adhesion with increased nanophase titania dispersion in PLGA. That is, when sonication power increased, ceramic agglomeration decreased which promoted titania dispersion and subsequently enhanced osteoblast adhesion. It is intriguing to consider why osteoblast adhesion was promoted on PLGA composites with well-dispersed titania. Increased surface coverage of nanophase titania particles and subsequently increased nanometer roughness may be plausible explanations. In fact, direct evidence is provided by this study that osteoblasts prefer titania over PLGA. Thus, as demonstrated in SEM pictures, osteoblast adhesion may have been enhanced simply because more titania was present on the surfaces of composites under high power sonication. In addition, previous studies have shown that protein interactions are much different on nanometer surfaces compared to conventional surfaces. Specifically, adsorption of vitronectin is much greater on bulk nanophase compared to conventional titania. Moreover, exposure of select epitopes that mediate osteoblast adhesion (such as RGD) was greater when vitronectin was adsorbed on nanometer compared to conventional ceramics. The same events may be happening here. Future studies will have to determine why osteoblast adhesion is promoted on PLGA composites with well-dispersed titania. Another explanation for increased osteoblast adhesion may be greater surface area. Previous studies have shown that, compared with larger grain size titania compacts, nanophase titania had about 35% more surface area for cell adhesion.[24] However, when normalized to this increased surface area, osteoblast adhesion, proliferation, and deposition of calcium-containing minerals were still enhanced on nanometer compared to conventional titania.[24] This indicates that increased surface area was not the contributing factor to greater osteoblast functions on nanophase titania compacts. The same may be true here, but still needs to be determined. Whatever the reason, this study suggests that nanophase titania in PLGA composites should be further considered in bone tissue engineering applications.

ACKNOWLEDGMENTS

The authors would like to thank the NSF for a Nanoscale Exploration Research (NER) grant and Ms. Janice Mckenzie, Dr. Rachel Price, Ms. Michiko Sato and Mr. Derick Miller for help with various aspects of this study.

REFERENCES

[1] A. R. Boccaccini and V. Maquet, "Bioresorbable and Bioactive Polymer / Bioglass® Composites with Tailored Pore Structure for Tissue Engineering Applications," *Composite Science and Technology*, **63**, 2417-2429 (2003).

[2] D. W. Hutmacher, "Scaffolds in Tissue Engineering Bone and Cartilage," *Biomaterials*, **21**, 2529-2543 (2000).

[3] R.C. Thomson, M.J. Yaszemski, J.M. Powers, and A.G. Mikos, "Hydroxyapatite Fiber Reinforced Poly(α-hydroxy ester) Foams for Bone Regeneration," *Biomaterials*, **19**, 1935-1943 (1998).

[4] R. Landers, U. Huebner, R. Schmelzeisen, and R. Muelhaupt, "Rapid Prototyping of Scaffolds Derived from Thermoreversible Hydrogels and Tailored for Applications in Tissue Engineering," *Biomaterials*, **23**, 4437-4447 (2002).

[5] J. Weng and M. Wang, "Producing Chitin Scaffolds with Controlled Pore Size and Interconnectivity for Tissue Engineering," *Journal of Materials Science Letters*, **20**, 1401-1403 (2001).

[6] Y. Zhang and M. Zhang, "Synthesis and Characterization of Macroporous Chitosan/Calcium Phosphate Composite Scaffolds for Tissue Engineering," *Journal of Biomedical Materials Research*, **55**, 304-312 (2001).

[7] M.A. Slivka, N.C. Leatherbury, K. Kieswetter, and G. Niederauer, "Porous, Resorbable, Fibre-reinforced Scaffolds Tailored for Articular Cartilage Repair", *Tissue Engineering*, **7** [6] 767-780 (2001).

[8] R.J. Vance, D.C. Miller, A. Thapa, K.M. Haberstroh and T.J. Webster, "Decreased Fibroblast Cell Density on Chemically Degraded Poly-lactic-co-glycolic Acid, Polyurethane and Polycaprolactone," *Biomaterials*, **25**, 2095-2103 (2004).

[9] H. Kim, H.W. Kim, and H. Suh, "Sustained Release of Ascorbate-2-phosphate and Dexamethasone from Porous PLGA Scaffolds for Bone Tissue Engineering Using Mesenchymal Stem Cells," *Biomaterials*, **24**, 4671-4679 (2003).

[10]V. Maquet, A.R. Boccaccini, L. Pravata, I. Notingher and R. Jerome, "Porous Poly(α-hydroxyacid)/Bioglass® Composite Scaffolds for Bone Tissue Engineering. I: Preparation and *in vitro* Characterization," *Biomaterials*, **25**, 4185-4194 (2004).

[11]W.W. Lu, F. Zhao, K.D.K. Luk, and Y.J. Yin, et al. "Controllable Porosity Hydroxyapatite Ceramics as Spine Cage: Fabrication and Properties Evaluation," *Journal of Materials Science: Materials in Medicine*, **14**, 1039-1046 (2003).

[12]P.X. Ma, R. Zhang, G, Xiao and R. Franceschi, "Engineering New Bone Tissue *in vitro* on Highly Porous Poly(α-hydroxyl acids)/hydroxyapatite Composite Scaffolds," *Journal of Biomedical Materials Research*, **54** [2] 284-293 (2001).

[13]H. Petite, V. Viateau, W. Bensaid, and A. Meunier, et al. "Tissue-engineered Bone Regeneration," *Natural Biotechnology*, **18** [9] 959-963 (2000).

[14]A.G. Stamboulis, L.L. Hench, and A.R. Boccaccini, "Mechanical Properties of Biodegradable Polymer Sutures Coated with Bioactive Glass," *Journal of Materials Science: Materials in Medicine*, **13** [9] 843-848 (2002).

[15]J. J. Blaker, J. E. Gough, V. Maquet, I. Notingher, and A. R. Boccaccini, "*In vitro* Evaluation of Novel Bioactive Composites Based on Bioglass® -filled Polylactide Foams for Bone Tissue Engineering Scaffolds," *Journal of Biomedical Materials Research - Part A*, **67** [4] 1401-1411 (2003).

[16]K.G. Marra, J.W. Szem, P.N. Kumta, P.A. DiMilla, and L.E. Weiss, "*In vitro* Analysis of Biodegradable Polymer Blend/hydroxyapatite Composites for Bone Tissue Engineering," *Journal of Biomedical Materials Research*, **47** [3] 324-335 (1999).

[17]S.J. Kalita, S. Bose, H.L. Hosick, and A. Bandyopadhyay, "Development of Controlled Porosity Polymer-ceramic Composite Scaffolds via Fused Deposition Modeling," *Materials Science and Engineering C*, **23** [5] 611-620 (2003).

[18]A.R. Boccaccini, J.A. Roether, L.L. Hench, V. Maquet, and R. Jerome, "A Composite Approach to Tissue Engineering," *Ceramic Engineering and Science Proceedings*, **23** [4] 805-816 (2002).

[19]A.J. Dulgar Tulloch, R. Bizios, and R.W. Siegel, "Nanophase Alumina/poly(L-lactic acid) Composite Scaffolds for Biomedical Applications," *Materials Research Society Symposium Proceedings*, **740**, 161-166 (2003).

[20]S. Kalita, J. Finley, S. Bose, H. Hosick and A. Bandyopadhyay, "Development of Porous Polymer-ceramic Composites as Bone Grafts," *Materials Research Society Symposium Proceedings*, **726**, 91-96 (2002).

[21]T.J. Webster, R.W. Siegel, and R. Bizios, "Osteoblasts Adhesion on Nanophase Ceramics," *Biomaterials*, **20**, 1221-1277 (1999).

[22]T.J. Webster, C. Ergun, R.H. Doremus, and R.W. Siegel, "Enhanced Functions of Osteoblasts on Nanophase Ceramics," *Biomaterials*, **21**, 1803-1810 (2000).

[23]T.J. Webster, R.W. Siegel, and R. Bizios, "Design and Evaluation of Nanophase Alumina for Orthopedic/dental Applications," *NanoStructured Materials*, **12** [5] 983-986 (1999).

[24]T.J. Webster and T.A. Smith, "Nano-biotechnology: The Design of More Effective Orthopedic Implant Materials Through the Use of Nanophase Ceramics," *Journal of Biomedical Materials Research*, in press, (2004).

NANOSTRUCTURED SENSOR MATERIALS FOR SELECTIVE BIO-CHEMICAL DETECTION

Arun K. Prasad and Pelagia I. Gouma
Dept. of Materials science & Engineering
SUNY at Stony Brook
Stony Brook, NY 11794, USA

ABSTRACT
There is a need for gas detectors to monitor inducers and products of biochemical processes; however the available technologies lack specificity. This work presents the sensing behavior of MoO_3-based films used in the detection of methanol and ammonia. Selectivity to these gases, in presence of each other and an interfering gas (CO) was demonstrated, and reproducibility of these results was studied. The microstructure of the sensing films controls the preferential sensitivity to the gases of interest compared to other vapors and the film thickness affects the sensitivity values for the specific gas detected. Future studies will address the issue of differential sensor selectivity within certain "families" of gases.

INTRODUCTION

Gas sensors have been used in a variety of applications. These include the monitoring of the concentration of toxic chemicals and their effect on the human health in industries; for pollution measurement and control; the detection and monitoring of explosive gas hazards; the control of gas and oil fired boilers in industrial, commercial and domestic premises; and more recently in the medical field for monitoring the chemical species in the blood and other body fluids.

A resistive gas sensor is a kind of chemical detector in which the changes in gas concentration are manifested as a change in resistance of the sensor material. The sensing elements are semiconducting metal oxides. MoO_3 and WO_3 were found in our previous work to be sensitive to a variety of gaseous species when tested at elevated temperatures (400 °C-500 °C).[1-3] It was also shown that the processing and microstructural properties greatly affect the sensing behavior of these binary oxides. A correlation was made between the "family" of gases to be detected and the crystallographic phase of the metal oxide that is sensitive to this particular family.[4] In this way, an informed design for novel gas sensor development is feasible.

The present work focuses on gas sensors as bio-detection probes for the monitoring of the concentration of gaseous metabolites that are important to biotechnology and medical diagnostics. The emphasis is given on the analysis of methanol and ammonia, since these are often the inducers[5] or products[6] of biochemical reactions. The aim is to develop detectors that are intrinsically selective to each of these two gases so as to eliminate interference from other volatile compounds. In contrast to the commercially available SnO_2 detectors that are sensitive to a broad range of gaseous species, the sensors developed in this work are structurally modified for inherent specificity to the gases of interest without the use of dopants or other secondary additions.

EXPERIMENTAL METHOD

Sol-gel processing is the process of preparing a sol, gelation of sol, and removal of solvent.[5] A sol is a colloidal suspension of solid particles in a liquid. A gel is a substance that contains a continuous solid phase enclosing a continuous liquid phase. The continuity of the solid phase gives the elasticity to the gel. The sol may be produced from inorganic or organic precursors. Alkoxides are the most common precursors used. Metal alkoxides have an organic ligand attached to the metal atom or metalloid atom. These alkoxides are used mainly because they react readily with water.[6]

$$M(OR)_z + H_2O \longrightarrow HO-M(OR)_{z-1} + ROH$$

where R is the alkyl group, and z depends on the valency of metal M.

Gels of molybdenum and tungsten oxides were prepared from molybdenum iso-propoxide and tungsten iso-propoxide precursors, respectively. These alkoxides were mixed with n-butanol to obtain 0.1 M solutions. After mixing, the sols were mechanically agitated for 5 minutes and then sealed airtight. Ultrasonic agitation was then performed for 2 hours and the sols were allowed to age and settle. In the case of molybdenum oxide sol, a black opaque liquid was obtained after 24 hours of aging. A transparent brownish yellow sol was obtained in the case of tungsten oxide sol after the same aging period.

Thin film was deposited by spin coating method and/or by dropping controlled amounts of the sol-gel on the sensor substrates. The spin coating was carried out at 1500 rpm for 30 sec and was repeated 5 times to obtain a film thickness of 150-175 nm. The films were left overnight for gelation and condensation and then sintered at 500 °C for 8 hour in air. This procedure was repeated another time to obtain the second set of films of thickness in the range of 250-275 nm. The thickness was measured after heat treatment in both cases. Alumina substrates equipped with a Pt heater on the backside and interdigitated Pt contacts on the top. These substrates were kindly provided to us by Prof. G. Sberbeglieri, of the Sensor Lab, Univ. of Brescia, in Italy.

Electrical measurements assessing the gas sensing response of individual elements were carried out in the gas flow bench at the author's lab at SUNY-Stony Brook, which is described elsewhere. The sensing response of arrays consisting of two elements respectively was studied using the one-of-a-kind facility at the Univ. of Brescia. Electrical measurements were conducted under a controlled humidity environment (10% RH). The temperature of the gas chamber was set at 20 °C. Sensing tests were carried out at operating temperatures in the range of 400-500 °C. The formula used to calculate sensitivity and response time data from these measurements is:

$$S=(R_g-R_{air})/R_{air}$$

where subscripts g and air indicate gas pulse and background oxygen respectively.

Structural characterization consisting of transmission electron microscopy imaging and electron diffraction studies was performed using a Philips CM12 transmission electron microscope with LaB_6 cathode and incident energy of electrons of 120 keV at SUNY-Stony Brook. For TEM imaging the thin films were deposited on holey copper grids that were heat-treated using similar conditions with those used for gas sensing tests.

RESULTS

Figure 1 shows the response of MoO₃ thin films prepared from sol-gel synthesis to different gases at 400 °C. The plot presents two sensing films of different thicknesses and their relative sensitivities to methanol (400 ppm), ammonia (15 ppm) and CO (15 ppm) at regular time intervals. The measurements were repeated every 5 hours. The sensitivity towards methanol remained stable around 0.67 and 0.75 respectively (as a function of increasing film thickness) during the 15 hours of testing. The sensitivity towards ammonia, however, showed an irregular trend but averaged at around 0.18. The sensor showed negligible sensitivity towards CO. When mixtures of methanol and the other gases were tested, the sensor responses remained close to those for single methanol measurements.

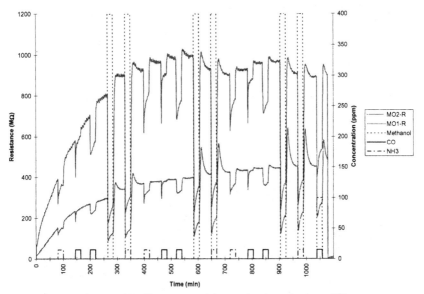

Figure 1. Response of MoO₃ thin films prepared from sol-gel synthesis to different gases at 400 °C.

Figure 2 shows the response of Mo-W mixed oxide sensing films tested at 500 °C to the same set of gases. The methanol sensitivity was considerably reduced (averaged at 0.27) while the ammonia sensitivity was raised to the same level (0.25).

Figure 3 shows the SEM image of MoO₃ films heat treated at 450 °C. The features shown evolving from the bulk indicate the growth of new grains (of the monoclinic phase). The film is highly porous with pore size in the range of 200-500 nm. In Figure 4, the microstructure of the film has evolved to well-oriented columnar crystals after heat treatment at 500 °C.

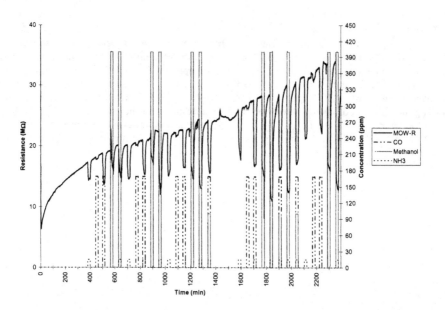

Figure 2. Response of (Mo-W)O₃ thin film prepared from sol-gel synthesis to different gases at 500 °C.

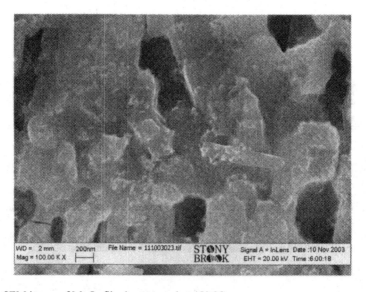

Figure 3. SEM image of MoO₃ film heat treated at 450 °C.

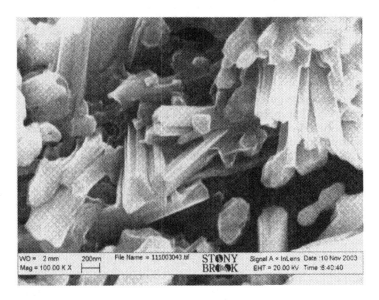

Figure 4. SEM image of MoO₃ film heat treated at 500 °C.

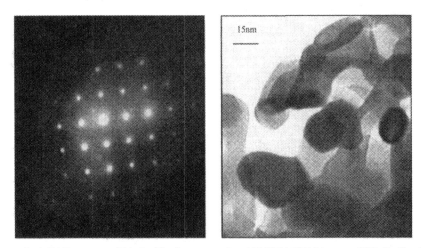

Figure 5. a) SAD pattern of MoO₃ film heat treated at 450 °C b) TEM image of MoO₃ film heat treated at 450 °C.

Figure 5 shows the selected area electron diffraction pattern from one of the grains of MoO_3 film after heat treatment at 450 °C oriented along the [010] zone axis of monoclinic phase (Space Group :$P2_1/c$ (14)) along with the corresponding micrograph of the grain structure and morphology. The average grain size was measured to be 20-30 nm.

DSC/TGA analyses performed on the sol-gel dried powders of MoO_3 (see Figure 6) indicated that at around 415 °C there is a formation of a new phase in MoO_3 from the as-received state which corresponds to the formation of the β–phase of MoO_3, the monoclinic phase, as is this is confirmed through XRD analysis. At around 475 °C, there is another peak suggesting that the β–phase undergoes transformation to the stable α-phase (orthorhombic) at this time.

XRD studies performed on these samples confirmed the phase transition. Figure 7 shows the XRD spectra of the two films heat treated at 400 °C and 500 °C with the major peaks identified. The presence of monoclinic phase in films heat treated at 400 °C is shown by the (101) peak. The presence of orthorhombic phase in films heat treated at 500 °C is evident by the appearance of strong (002) peak, which is almost absent at 400 °C.

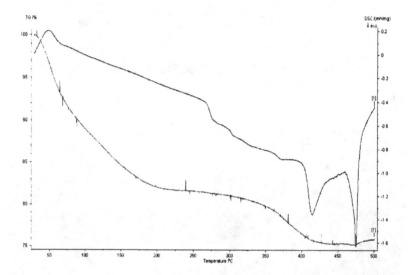

Figure 6. DSC/TGA study of MoO_3 sol-gel dried powders showing peaks at 415 °C and 475 °C.

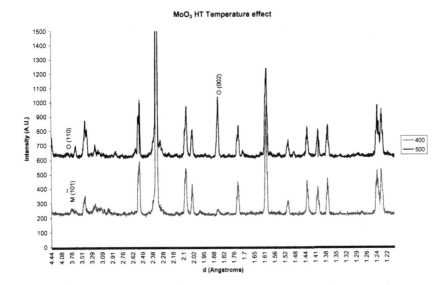

Figure 7. XRD of sol-gel films heat treated at 400 °C and 500 °C.

DISCUSSION

Numerous studies of metal oxides systems for gas sensors have approached the issue of gas discrimination (selectivity) by adding secondary elements/phases or altering the operation temperature of the detector. The argument made here is that the microstructure and crystallography of the sensor material are the key parameters controlling selectivity. It is shown in this work that the thickness (150 nm & 250 nm) of the oxide film plays a role in enhancing the sensitivity to the gas detected but does not have any effect on the shape of the signal (change of resistance with concentration during insertion and removal of the gas pulse) or the relative sensor sensitivity with respect to interfering gases.

The deciding factor whether the oxide will respond strongly to methanol versus ammonia was shown to be the phase of the material stabilized at the temperature of operation. The thermal analyses results, complemented by the structural investigations, have determined the phase fields for the sol-gel-processed films of the MoO_3 system and there is a clear phase transition occurring in the films between the two testing conditions used in this study. According to the model proposed earlier,[4] ammonia will be selectively detected by the open structure of the orthorhombic polymorph, whereas alcohol-related affinity characterizes more tetragonal structures. Future studies will focus on assessing the relative specificity of these sensors to various members of families of gases, thus testing for various alcohols and amines is underway.

CONCLUSIONS

The gas sensing behavior of MoO_3 thin film sensors was assessed by varying the processing conditions namely thickness and stabilization heat treatment temperature. The increase in thickness merely has an effect in increased sensitivity to the gases. Selectivity to methanol was demonstrated for films tested at 400 °C, in the presence of CO and ammonia. The sensitivity towards ammonia was improved in the presence of methanol and CO by performing the sensing tests at 500 °C. Phase structure and morphology are key factors to be considered for selective chemical detection. The effect of morphology on sensing behavior is under study and will be reported in future communication. The gases tested are of major importance to biotechnology applications.

ACKNOWLEDGEMENTS

This work has been funded by the National Science Foundation (NIRT grant with Dr. L. Madsen as program manager) and a WISC grant awarded to PG.

REFERENCES

[1]A.K. Prasad, D. Kubinski, and P. I. Gouma, "Comparison of Sol-Gel and RF Sputtered MoO_3 Thin Film Gas Sensors for Selective Ammonia Detection," *Sensors & Actuators B*, **9**, 25-30 (2003).

[2]A.K. Prasad, P.I. Gouma, D. J. Kubinksi, J.H. Visser, R.E. Soltis, and P.J. Schmitz, "Reactively Sputtered MoO_3 films for ammonia sensing," *Thin Solid Films*, **436**, 46-51 (2003).

[3]A.K. Prasad, P.I. Gouma, "MoO_3 and WO_3 based thin film conductimetric sensors for automotive applications," *J. Mat. Sci.*, **38** [21] 4347-4352 (2003).

[4]P. I. Gouma, "Nanostructured Polymorphic Oxides for Advanced Chemosensors," *Rev. Adv. Mater. Sci.*, **5**, 123-138 (2003).

[5]M.M Guarna, G.J. Lesnicki, et al, "On-line Monitoring and Control of Methanol Concentration in Shake-Flask Cultures of Pichia pastoris," *Biotechnology and Bioengineering*, **56** [3] 279-286 (1997).

[6]U. Bilitewski, W. Drewes, and R.D. Schmid, "Thick-Film Biosensors for Urea," *Sensors and Actuators B-Chemical*, **7** [1-3] 321-326 (1992).

INVESTIGATION OF ECOLOGICAL SAFETY IN USING NANOSIZED AND ULTRAFINE POWDERS

I. Uvarova, N. Boshitskaya, V. Lavrenko and G. Makarenko
I.N. Frantsevich Institute for Problems of Materials Science, NAS of Ukraine
3 Krzhizhanovsky Str.
03142 Kyiv, Ukraine

ABSTRACT

High dispersity of nanosized and ultrafine powders used for nanostructured material preparation requires investigation of their behavior in different media, in particular at the contact with the main fluids of living organism, such as blood plasma, digestive juices and physiological solution. The behavior of different nanosized silicon, aluminum, boron and titanium nitrides prepared by different methods in solutions, which imitate blood plasma, digestive juices and physiological solution, has been discussed.

Of different aluminum nitride materials studied in this work, filamentary crystals had the highest stability in biochemical media. Aluminum nitride ceramic was established to be ecologically safer than silicon nitrides. Titanium nitride showed the highest stability in all media. Its use in stomatology instead of stainless steel is very promising thanks to its inactivity in interaction with digestive juice.

The data obtained give us possible evidence of using physicochemical investigations instead of direct tests on animals.

INTRODUCTION

Health and Safety are serious issues during the development of modern technology of powder metallurgy, especially in production of ultrafine and nanosized powders. Safety of working with powders of different chemicals and their dispersity are usually determined by their action on animals. To reduce the cost and time of investigations and to decrease the number of experiments on animals, novel method of physicochemical study of the behavior of different powder materials in biochemical media was proposed instead of direct experiments on animals. Knowledge of refractory powder compounds interaction with biochemical media such as digestive juices, saliva and blood plasma is a very important factor for their applications in medicine, and for estimation of their safety during ultrafine powder production since their dusts may penetrate into mouth cavity, lungs and stomach. Today various ceramic materials with high wear and corrosion, high-temperature and abrasion resistances are widely used instead of metals. Ecology of production and surrounding medium is determined by toxicity of used powder materials and especially by their dispersity and particle morphology. Investigations of powder behavior in blood plasma permit one to make prognosis for a real influence of toxic materials on stomach and lungs. The structure, morphology and dispersity of refractory powders are determined by the methods of their production.

The aim of this work is to determine the influence of powder dispersity, structure and types of chemical bonds in silicon, aluminum, boron and titanium nitrides on the peculiarities of powder interaction with biochemical and inorganic media for forecasting chemical and biochemical processes in human organism which take place after ingress of these powders into it. The interaction of nitrides with biochemical media, such as blood plasma, digestive juices and physiological solution, and also with inorganic media such as distilled water, solutions of HCl, KOH, NaCl with the same pH, was studied. The comparative investigations of behavior of titanium nitride and stainless steel in water, saliva, digestive juices and salt solution were carried out to estimate their toxicological action while using coronas and dentures.

EXPERIMENTAL

To investigate the properties of refractory powder materials, the following biochemical media were used: Ringer-Lock physiological solution, close to blood plasma by composition and osmotic pressure, blood serum of cattle, and digestive juices. As control media, a distilled water, and aqueous solution of HCl (pH =2.0), and NaOH (pH =7.4) were used.

The control media were chosen in accordance with closeness of their acidity to that of digestive juice and blood plasma.

Chemical composition was determined by the nitrogen content in the initial powders and in the insoluble residue and in filtrate. X-ray diffraction analysis and transmission electron microscopy were used for structure, morphology and phase composition determination.

RESULTS AND DISCUSSION

Si_3N_4 System

Silicon nitride has two modifications (low-temperature - α and high-temperature - β); both of them form at the same time in such processes as plasma chemical and high temperature self-propagating synthesis. Silicon nitride has moderate poisonous action.[1,2] During inhalation exposure of technical Si_3N_4 powder (nitriding of silicon in nitrogen flow) on laboratory animals, small changes in biochemical indices in lungs, stomach and blood were detected. Morphological picture of lungs tissue was characterized by thickening the vessel walls, accumulating powder particles, expanding conjunctive tissue and a small diffusion sclerosis. Ultrafine and nanosized Si_3N_4 powders with a particle size of about 40 nm have 7 times greater toxicity on the human organism. Nanosized Si_3N_4 powders having unique properties such as wear, corrosion and heat resistance, and high hardness have shown wide applications. Si_3N_4 ceramics are used in biology[3] and biomedicine[4] and as coatings on medicine instruments and implants.[5,6]

Si_3N_4 has 78 % α-Si_3N_4 phase and a mean size of particles of 4-5 μm.[7-9] Carbon-thermal reduction results in fibrous powder formation. Nanosized Si_3N_4 powders prepared by the plasma chemical method are agglomerated in a form similar to that of technical powders but these agglomerates consist of fibers.

Interaction of all these powders with physiological solution, distilled water, digestive juice, diluted HCl and NaOH solutions and blood serum is shown in Tables I and II. The technical Si_3N_4 is stable in diluted solutions of HCl and NaOH. Such powders slightly dissolve in distilled water, digestive juice and in blood serum. The interaction with physiological solution is more intensive. The fibrous powders are more stable in different media except blood serum. Plasma powders are the most active in all media, especially in physiological solution and blood serum. Of the media investigated, physiological solution is the most active solvent. Dissolution of Si_3N_4 in digestive juice insignificantly depends on the morphology of powders.

X-ray diffraction analysis was carried out on all powders before and after the interaction with digestive juice and blood serum. All initial powders contained α- and β- phases of Si_3N_4 and a small amount of oxynitride phase.

Table I. Content of nitrogen in Si$_3$N$_4$ samples after interaction with biochemical and inorganic media.

Type of Si$_3$N$_4$ powder	Initial	N (mass %)					
		Neutral medium		Acid medium		Alkaline medium	
		Physiological solution	H$_2$O (pH 7.0)	Digestive juice	HCl (pH 2.0)	Blood serum	NaOH (pH 7.4)
Technical	38.5±0.17	27.7±0.2	35.5±0.13	34.5±0.18	38.6±0.2	33.2±0.17	38.0±0.19
Fiber	33.1±0.17	31.1±0.22	31.7±0.11	29.8±0.16	32.6±0.16	24.0±0.2	32.8±0.25
Nanopowder	38.6±0.11	5.2±0.23	32.5±0.23	33.8±0.14	35.6±0.11	12.5±0.18	34.17±0.1

Table II. Interaction of Si$_3$N$_4$ samples with neutral, acid and alkaline media

Type of Si$_3$N$_4$ powder	Transformation degree, α (%)					
	Neutral medium		Acid medium		Alkaline medium	
	Physiological solution	H$_2$O (pH 7.0)	Digestive juice	HCl (pH 2.0)	Blood serum	NaOH (pH 7.4)
Technical	28.0	7.8	12.2	0	13.8	1.3
Fiber	6.0	4.0	10.0	1.45	26.3	0.8
Nanopowder	86.5	16.9	13.7	7.8	67.6	11.5

Silicon nitride prepared by carbon thermal reduction is a mixture of monocrystalline fibers, crystallites and amorphous phase. The main lines are α-Si$_3$N$_4$. Oxynitride is practically absent. There is a small mixture of β- Si$_3$N$_4$ phase. Like the technical Si$_3$N$_4$, the fibrous silicon nitride keeps its phase composition during interaction with physiological solution, diluted HCl and NaOH solutions and water. Fibrous silicon nitride interacts with blood serum in the greatest degree. X-ray diffraction line intensities decrease by 1.5 times, whereas the phase composition does not change.

Thus, in both cases proteins of blood serum accelerate the process of interaction of silicon nitride with medium.

The initial silicon nitride powder prepared by the plasma chemical method has crystallites of 0.02 μm in size, mostly in an agglomerated state. As a result of plasma powder interaction with physiological solution, the content of nitrogen in the solid residual decreases to 5.2±0.23 and transformation degree is very high at 86.5% (Tables I and II). The intensity of the β-Si$_3$N$_4$ diffraction lines decreases by 1.5 times, but the intensity of the α-Si$_3$N$_4$ diffraction lines decreases slightly. The dissolution of nanopowder in distilled water is 5 times slower and the transformation degree is 16.9 % (Tables I and II). The content of nitrogen in the solid residual after interaction with digestive juice decreases to 33.8±0.14 with the transformation degree equal to 13.7 %. The ratio between α- and β- Si$_3$N$_4$ practically does not change in comparison with the initial powders. Thus, organic compounds of digestive juice stimulate the dissolution of silicon nitride in acid media.

In the Si$_3$N$_4$ nanopowder - blood serum system the content of nitrogen in the solid residual decreases to 12.5±0.18 and the transformation degree is 11.5 % (Tables I and II). The solid phase is characterized by the presence of α- and β- Si$_3$N$_4$ lines, but their intensities are smaller.

Thus, silicon nitride nanopowder interacts with physiological solution and blood serum in the greatest extend. In the control inorganic media it is stable enough.

The chemical analysis of the filtrate after interaction of technical, fibrous and plasma silicon nitrides with biochemical media showed very small contents of nitrogen and silicon (Table III). Such a small amount of nitrogen in the filtrate may be connected with outlet of it in the molecular form, with silicon remaining in the insoluble residue as a gel of silicic acid.

Table III. Results of chemical analysis of filtrate after interaction of silicon nitride with biochemical media

Type of Si_3N_4 powder	Physiological solution		Digestive juice		Blood serum	
	nitrogen	silicon	nitrogen	silicon	nitrogen	silicon
Technical	1.4	0.055	2.2	0.21	-	0.73
Fiber	2.8	0.069	2.2	0.14	-	0.67
Nanopowder	0.8	0.189	2.1	0.22	-	0.69

Nanosized silicon nitride powder can be found in any biochemical and control media as a colloid system consisting of micelles formed as a result of interaction of ultrafine phase with electrolyte.

Thus, the investigation of silicon nitride interaction with different media has shown that the most active biochemical media are blood serum and physiological solution, which can be connected with the reaction acceleration thanks to the presence of albuminous ferment clusters (in bioinorganic media) or of catalytic active Ca^{2+} ions (in physiological solution).

The data obtained were confirmed by the results of microscopic investigations of rat lungs tissues after injection into lungs of technical silicon nitride which shows swelling of interalveolar membranes.[8] The intensive cellular infiltration took place too. The epithelium of lungs did not change and collagen formation was not detected. Such data can be explained as an upcoming dust disease in diffusion- sclerotic form.

Thus, the presence of local collagen neoplasm areas results in fibrogenic action of nanosized silicon nitride powders on organism of white rats. Technical silicon nitride powder has a small toxic, allergen and fibrogenic action on white rats. Fibrous silicon nitride has a more intensive toxic action. Nanocrystalline silicon nitride has the most toxic action and can initiate considerable changes in cells.

AlN System

AlN^{10} (reduction and nitriding of alumina) contains agglomerates with a mean size of 20-60 μm, and filamentary AlN crystals consist of near 90 mass % fibers (ratio of the fiber length to diameter is equal to 10:1) and 10 mass % powder particles with a size of 3-5 μm. In both cases X-ray diffraction analysis indicated lines of only the AlN phase with a small displacement to smaller lattice parameters. It may be connected with dissolution of small amount of oxygen in AlN. Nanosized AlN powders agglomerated to particles with a size range between 100 nm and 500 nm. X-ray diffraction analysis of nanosized AlN powders prepared by the plasma chemical method also showed the main lines of the AlN phase and possible lines of the AlO_xN_y as an admixture. Plasma chemical powder hydrolyzes very easily even in air.

Interaction of all these powders with physiological solution, distilled water, digestive juice, diluted HCl and NaOH solution and blood serum is shown in Tables 4 and 5. Technical AlN is hydrolyzed in water. The content of nitrogen in the solid residue decrease to 20.8±0.1 mass% and the transformation degree is 32.5%. According to the X-ray diffraction data, the β-$Al_2O_3\cdot3H_2O$ oxide phase forms in this case. Its interaction with physiological solution is similar. The AlN lines move to smaller lattice parameters and their intensities decrease twofold. In this case, the β-$Al_2O_3\cdot3H_2O$ oxide phase is present too. The content of nitrogen in solid residue decreases twofold down to 20.1±0.1 mass% and the transformation degree is 34.7 %. As a result of the interactions in the aluminum nitride - blood serum system, the content of nitrogen in solid residue decreases down to 24.9±0.1 mass% and the transformation degree is 19.1 %. According to the X-ray diffraction data, herein only the AlN phase is present. The intensity of its line decreases a little. Other lines were not found. The content of nitrogen in solid residue after interaction with digestive juice is 26.5±0.1 mass% and the transformation degree is 14.0 %. Diffraction lines are the same as those from the initial powder. It can be supposed that the presence of different organic acids, proteins, lipids and also ferments of protease and lipase in digestive juice promotes the formation of complexes with decreasing the reactivity of media.

Table IV. Content of nitrogen in AlN samples after interaction with biochemical and control media

Type of AlN powder	N (mass %)						
	Initial	Neutral medium		Acid medium		Alkaline medium	
		Physiologic al solution	H_2O (pH 7.0)	Digestive juice	HCl (pH 2.0)	Blood serum	NaOH (pH 7.4)
Technical	30.8±0.2	20.1±0.1	20.8±0.1	26.5±0.4	4.3±0.1	24.9±0.2	7.3±0.2
Fiber	32.9±0.2	29.3±0.2	27.5±0.2	29.9±0.1	1.5±0.1	32.7±0.1	2.4±0.1
Nanopowder	26.5±0.2	15.1±0.2	12.5±0.2	1.27±0.3	1.7±0.1	13.2±0.3	0.35±0.1

In the system of filamentous AlN crystal - physiological solution, the partial dissolution of aluminum nitride is observed. The intensity of AlN lines decreases and traces of the β-$Al_2O_3 \cdot 3H_2O$ phase appear. The content of nitrogen in solid residual after interaction with physiological solution is 29.3 mass% against 32.9 mass% in the initial sample. The transformation degree is 10.9% (Tables IV and V). Thus, filamentous AlN interacts with a physiological solution worse than technical AlN. Filamentous AlN interacts with water better compared to physiological solution but worse than technical AlN. The content of nitrogen in solid residual decreases to 27.5 mass% and the transformation degree is 16.4 %. Filamentous AlN practically does not interact with blood serum (transformation degree is 0.6%). The transformation degree is 9.1 % during interaction with digestive juice. Chemical activity of filamentous AlN with respect to HCl and NaOH was established.

Thus, filamentous AlN crystals are more stable in biological media than technical AlN, which can be connected with their more ordered structure, close to the monocrystalline one.

Table V. Interaction of AlN samples with biochemical and inorganic media

Type of AlN powder	Transformation degree, α (%)					
	Neutral medium		Acid medium		Alkaline medium	
	Physiological solution	H_2O (pH 7.0)	Digestive juice	HCl (pH 2.0)	Blood serum	NaOH (pH 7.4)
Technical	34.7	32.5	14.1	86.0	19.1	76.2
Fiber	10.9	16.4	9.1	95.4	0.6	92.7
Nanopowder	43.0	52.8	95.2	93.6	50.1	98.6

Plasma chemical powder is the most active in all the media. It decomposes in water and forms Al(OH)₃. The decomposition degree is 52.8 % (Table V). In physiological solution, the decomposition is not so active; the decomposition degree equals to 43 % and the content of nitrogen in solid residual is 15.1 mass%. The X-ray diffraction analysis indicates to a decrease in AlN line intensity by three times and to the presence of the β-$Al_2O_3 \cdot 3H_2O$ phase. Nanosized AlN powder interacts very actively with digestive juice (transformation degree 95.2 %), blood serum (transformation degree 50.1 %) and especially with control media such as HCl (transformation degree 93.6 %) and NaOH (transformation degree 98.6 %). The β-$Al_2O_3 \cdot 3H_2O$ phase is present in all cases of interaction of plasma chemical powder with biochemical and control media.

Thus, comparison of AlN powders of different dispersity shows an important role of the specific surface area and structure of materials in their behavior in biochemical media. The decrease in the rate of AlN interaction with biochemical media compared to inorganic ones can be attributed to the inhibitory action of organic components especially those present in blood serum.

TiN System

There are two types of TiN powders: prepared by titanium nitration and plasma chemical methods. The stability of these powders in different media is shown in Table IV.

As can be seen that TiN powders are the most stable in different media in comparison with other systems above.

Now TiN is a well-known biological implant and it forms biological bonds with surrounding tissue within six weeks after implantation. The toxic reactions have not being observed.[11] To study the reaction of TiN powder on the living organism, intrapulmonary injection into rat's lungs was carried out. Only little changes in lungs tissues, liver and kidneys were observed. Our investigations established that the content of collagen albumens in rat's lung tissues after three months increased by 25.8 % (TiN, nitration) and 38.5 % (TiN, plasma). However, after six months a decreasing tendency in the collagen content was observed. The morphological picture of lung tissues after three and six months did not change.

Table VI. Interaction of TiN samples with biochemical and inorganic media

Method of TiN preparation	Transformation degree, α (%)					
	Neutral medium		Acid medium		Alkaline medium	
	Physiological solution	H_2O (pH 7.0)	Digestive juice	HCl (pH 2.0)	Blood serum	NaOH (pH 7.4)
Nitration	3.3	0.94	1.89	3.6	2.83	0.47
Plasma	4.0	1.1	1.92	4.5	2.96	0.97

BN System

Boron nitride, a compound with a primarily covalent type of chemical bonds, is very stable against the influence of different reagents. The stability of boron nitride in water, mineral acids and alkalis depends on its purity and crystal structure ordering. A wide use of boron nitride in techniques requires investigations of its toxic properties.

The data of microscopic investigations of BN powders after interaction with different biochemical media showed that the mean size of these powders does not change; only destruction of agglomerates has been observed.[12] X-ray diffraction analysis show very small changes after interaction of different types of BN with different biochemical media, which results in a small toxic action of BN which can be shown up only as an irritating factor because of the presence of powder in mouth cavity and lungs.

CONCLUSIONS

1. Study of the effect of different refractory compounds on biochemical media permits one to reduce the cost of experiments and decreases the number of experiments on animals.

2. Silicon nitride powders are rather active in physiological solution, blood serum, and digestive juice. The activity of these powders increases with an increase in their dispersity.

3. Inhibition effect of biochemical components has been established for the AlN system. The activity of AlN is higher in the control media than in the biochemical ones.

4. The BN system practically does not dissolve in biochemical media.

REFERENCES

[1]I.T. Brachnova, "Toxicity of metal powders and their compounds," Kiev, "Naukova dumka", 121, 1971.

[2]I.T. Brachnova, "Toxicity of metal powders used in powder metallurgy", Kiev, IPM NAS of Ukraine, 154, 1970.

[3]J.M. Buckiey-Golder, "Applications of ceramic films," pp. 33-45 in *Appl. Eng. Ceram.*, Engine Appl. Pap. Seminar, London, 1986.

[4]N. Terao, "Panorama of bioceramics realization," *Silicat ind.*, **52** [9] 123-28 (1987).

[5]Plant Martin, "Automated difractometry boosts laboratory productivity in ceramics research," *Ceram. Ind. J.*, **96** [1064] 90-92 (1987).

[6]R. Westerheide, "Ein werkstoff unter der lupe," *Schweiz. Maschinenmarkt*, [50] 48-50 (1995).

[7]N.V. Boshitskaya, T.S. Bartnitskaya, G.N. Makarenko, V.A. Lavrenko, N.M. Danilenko, N.P. Telnikova, "Chemical stability of Si_3N_4 in biochemical media," *Powder Metallurgy and Metal Ceramics,* **35** [9/10] 497-99 (1996).

[8]T.S. Bartnitskaya, N.V. Boshitskaya, V.A. Lavrenko, G.N. Makarenko, O.I. Popova, "Interaction of Si_3N_4 powders with media of living organism," *Ukrainian Chemical Journal,* **63** [1-2] 16-19 (1997).

[9]N.V. Boshitskaya, T.S. Bartnitskaya, V.A. Lavrenko, G.N. Makarenko, "Interaction of Si_3N_4 nanopowders with biochemical media," pp.235-36 in *Proc. VI Conference European Ceramic Society,* Brighton, UK, 1999.

[10]N.V. Boshitskaya, V.A. Lavrenko, T.S. Bartnitskaya, G.N. Makarenko, G.A. Chkurko, N.M. Danilenko, "Interaction of AlN powders with biochemical media," *Powder Metallurgy and Metal Ceramics,* **39** [3/4] 153-62 (2000).

[11]Mohanty Mira, K. Rathiman, C.C. Kartha, "Long Term Soft Tissue Response to Metals. Comparative Histopathological Evaluation," *Bull. Mater. Sci* [5] 309-15 (1987).

[12]N.G. Kakazey, V.A. Lavrenko, N.V. Boshitskaya, G.A. Chkurko, "Interaction of boron nitride powders with biochemical media," *Powder Metallurgy and Metal Ceramics,* **39** [1/2] 73-77 (2000).

PANEL DISCUSSION: NANOTECHNOLOGY - PAST, CURRENT, AND FUTURE

S. W. Lu
PPG Industries, Inc.
Glass Technology Center
P. O. Box 11472
Pittsburgh, PA 15238-0472

Summary of a panel discussion held in conjunction with Symposium 6, "Nanostructured Materials and Nanotechnology" during the 106th Annual Meeting of the American Ceramic Society, Wednesday, April 21, 2004, 11:00 am - 12:00 noon in Indianapolis, Indiana

INTRODUCTION

A significant increase of presentations and attendance in the Symposium 6 "Nanostructured Materials and Nanotechnology" during the 106th Annual Meeting and Exhibition of the American Ceramic Society (ACerS) over last year has showcased tremendous interests in the field of nanotechnology and ceramic nanomaterials from researchers, scientists and engineers all over the world. This year, the Symposium "Nanostructured Materials and Nanotechnology", sponsored by PPG Industries, Inc. and Basic Science Division of the American Ceramic Society, was the largest and most well-attended symposium during the three-day conference. In conjunction with the symposium, a panel discussion "Nanotechnology: Past, Current, and Future" was held on Wednesday, April 21, 2004, in Indiana Convention Center & RCA Dome, Indianapolis, Indiana. This is the second panel discussion in a series of ACerS symposia related to nanotechnology, following the first nanotechnology panel discussion "Commercialization of Nanomaterials" in conjunction with Symposium 7 "Nanostructured Materials and Nanotechnology" during the 105th Annual Meeting and Exhibition of the ACerS in Nashville, Tennessee, in 2003. This time, six professors and scientists who are involved with nanotechnology and nanostructured materials and their applications were invited as panelists to discuss important issues related to nanotechnology, its current research focuses, and future trends. The panel discussion was highlighted due to the fact that all panelists presented their recent research results during the Symposium 6 "Nanostructured Materials and Nanotechnology". Approximately eighty people attended and listened to the hour-long discussion and interacted with panelists during and after the discussion.

The six invited panelists were: James H. Adair, Professor of Materials Science and Engineering, Director, National Science Foundation (NSF) Particulate Materials Center at the Pennsylvania State University, PA; Mohan Edirisinghe, Professor of Materials, Queen Mary, University of London, UK; Kazuo Furuya, Director of Nano-characterization Research Group, Nanomaterials Laboratory, National Institute for Materials Science, Japan; John A. Rogers, Professor of Materials Science and Engineering, Founder Professor of Engineering, and Professor of Chemistry, University of Illinois at Urbana-Champaign, IL; Zhong Lin 'ZL' Wang, Regents' Professor, Director of Center for Nanoscience and Nanotechnology and Director of Center for Nanostructure Characterization, Georgia Institute of Technology, GA; Thomas J. Webster, Assistant Professor of Biomedical Engineering and Materials Engineering, Director of the Nanostructured Biomaterials Laboratory, Purdue University, IN. The panel discussion was organized and moderated by Dr. Song Wei Lu of PPG Industries, Inc., PA. All panelists introduced themselves to the audience before the panel discussion. The panelists discussed three pre-selected questions and three other questions from the audience during the panel discussion. Three questions were pre-selected by the symposium organizing committee (Song Wei Lu of PPG Industries, Inc., PA, Michael Hu of Oak Ridge National Laboratory, TN, and Yury Gogotsi of Drexel University, PA). Notes were taken during the panel discussion. The panelists were given the chance to review this summary for accuracy prior to publication.

TOP TEN RESEARCH AREAS IN CERAMIC NANOTECHNOLOGY THAT SHOW THE MOST SIGNIFICANT IMPACT TO OUR SOCIETY

The panel discussion started with the first pre-selected question "What are the top ten specific areas that ceramic nanotechnology show the most significant impact to our society as well as the best potential for commercialization?" The panelists listed the areas they thought were very important in the field of ceramic nanotechnology. Wang, a prominent professor in nanotechnology, and ranked as one of the most cited authors in nanotechnology from 1992 to 2002 by Science Watch, pointed out his choices of top nanotechnology research areas which will impact the society in the coming years: health, drug delivery, microelectronics, environment, energy, catalysis, and hydrogen-based materials. Wang investigates science and applications of nanoparticles, nanowires and nanobelts, functional oxide and smart materials for sensing and actuating, and nanomaterials for biomedical applications and nanodevices. Wang's group has recently applied the semiconducting nanobelt materials to make field effect transistor and single wire sensors. Adair, a leading scientist in dispersing and processing of nanoparticles, suggested top research areas as health, bio-imaging, drug delivery, gene therapy, electronics, multi-layered device, and varistor. Adair is the director of National Science Foundation (NSF) Particulate Materials Center at the Pennsylvania State University, dedicated to support member research and manufacturing interests by developing engineering and scientific foundations for the manufacture of advanced particulate materials (10 microns or less), including nanomaterials. Current research works include nano-sized α-alumina, synthesis, dispersion, characterization & self-assembly, synthesis of fluorescent nanocomposite particles, synthesis of fluorescent nanocomposite particles, aggregate breakdown and aqueous processing of ZnO based varistors, electrophoretic deposition studies of nanoparticles for thin films and coatings, absorption and light scattering of small particle pigments.

Edirisinghe explained the importance of the nanotechnology areas with foreseen time, i.e. pigments, coatings, catalysis, flat panel display, fuel cells, and cosmetics applications in four to ten years, and targeted drug delivery in twenty years. Edirisinghe, a professor of materials at Queen Mary, University of London, conducts researches on powder-based processing and modeling, in particular the application of aerosol science and technology to the forming and processing of structural, functional and biological materials, especially advanced inorganic materials. His recent research has also moved to form nanostructures by aerosol process [see J. Appl. Ceram. Tech., 1 (2004) 140-145]. Furuya emphasized about the significance of applications whose value added with nanomaterials. He listed few examples such as the combination of carbon nanotube with ceramics, or carbon nanotubes with metals. This point of view echoed the classification of nanomaterials applications to two categories during the panel discussion in 2003: nanomaterials themselves and products with value added by nanomaterials. Furuya's research interests spans from *in-situ* electron microscopy of engineering materials, to atomic level observation and analysis of materials to advanced electron and ion beam technology. He has reported the *in-situ* observation of nanocrystal agglomeration, merging and growth, and formation of nanomaterials under high-resolution transmission electron microscope.

Rogers added that areas such as bio-nanotechnology, bone implants, drug delivery, photonics, optical fibers by sol-gel routes, acoustic filters, wireless applications, high frequency applications, ultrahard materials such as SiC and diamonds, microelectronic and macroelectronic materials, thin films, aerogels, colloidal materials, and printable solar cells are very important. Rogers, a former division director at Bell Laboratories, Lucent Technology, and now Professor of Materials Science and Engineering, Founder Professor of Engineering, and Professor of Chemistry, at the University of Illinois at Urbana-Champaign, works on flexible paperlike displays, tunable microfluidic optical fiber, stamping techniques with nanometer resolution and high speed nematic liquid crystal microcell modulators built between the tips of optical fibers. His group developed new tools for fabricating structures with micron and nanometer dimensions that are critical to the progress of nanoscience and nanotechnology. His project seeks to develop soft lithographic methods for nanofabrication, and to use them for building structures that are needed for basic and applied studies.

Webster talked about areas such as those that integrate biology with nanotechnology to design better devices for the improvement of human health such as: environmental filters to clean contaminated environments, novel nanostructured materials in tissue engineering, enhanced functions of cells (such as bone, cartilage, vascular, bladder, and neuron) on nanoscale materials, increased bacteria attachment on ceramics, environment wastewater treatment, and drug delivery. He indicated that nano-sized hydroxyapatite is already commercialized and, thus, may represent one of the first successful commercialized nanophase products. Webster's research addresses the design, synthesis, and evaluation of nanophase (that is, materials with fundamental length scales less than 100 nm) materials as more effective biomedical implants. He has been at the forefront of investigating nanoscale materials in biological applications and has been awarded for such work through various awards such as the Biomedical Engineering Society Young Investigator Award (2002). He was the first to demonstrate an important size dependent relationship of increased new bone growth on nanograined materials like ceramics, metals, and polymers.

Combining all points of view from six panelists, the top specific areas in ceramic nanotechnology show the most significant impact to our society as well as the best potential for commercialization are: biological applications (improving implants, targeted drug delivery, bio-labeling, gene therapy, etc.), environmental applications (catalysis, waste water treatment, gas separation, etc.), energy applications (fuel cells and solar cells, hydrogen-based materials, etc.), optical and electronic applications (flat panel display, microelectronics, photonics, optical fibers, anti-reflective coatings etc.), cosmetic applications, chemical applications (pigments, coatings, etc.), and mechanical applications (hard coating), and others not yet discovered.

MANUFACTURABILITY OF NANOTECHNOLOGY

After few seconds of silence when the moderator Lu asked the attendees for a question to the panel, a good question emerged: What is the manufacturability of nanotechnology and how to scale up nanotechnology? This question is actually more related to the commercialization of nanotechnology, which was the central theme of the Industrial Track Program this year, and the topic of last year's nanotechnology panel discussion: commercialization of nanomaterials. Audiences are eager to know the possibility of nanomanufacturing, and always pondering its feasibility. The answer to this question is difficult since manufacturing of nanomaterials is application and product dependent. Rogers, a former Director of Condensed Matter Physics at Bell Laboratories, Lucent Technology, emphasized that the economic way to make use of nanomaterials is to build nanomaterials right in the place they are needed, i.e. carbon nanotubes on devices. Nanofabrication of growing nanowires on the devices remains a big research challenge. Rogers focused on soft lithographic methods for building 2D and 3D nanophotonic systems and for constructing organic transistors and diodes that have nanometer or molecular scale dimensions. Wang pointed out that some processes do not require large amount of materials, in which quality and purity might be more important. Depending on applications, nanofabrication is not measured by quantity, it is quality and purity. For example, in the case of single wire sensors and transducers and actuators using semiconducting nanobelts, it requires small amount of defect-free nanobelts with high quality and high purity. Webster added that it needs high quality of nanomaterials for biological applications, too. In the case of targeted drug delivery, the quantity required is very small. Industrial collaborations with academia in biological applications of nanomaterials are critical. In order to speed up the commercialization of nanomaterials for biological applications, the industries should have a clear vision of what is wanted and a solid financial support. Furuya agreed that quality is more important than quantity in some applications.

In contrast, Adair believes that dispersion of nanosized powders is very critical. Removal of solvent and organic is rather difficult after dispersion. It involves numerous processes to make nanoparticles and disperse nanomaterials, to manipulate the particle surface, and to remove solvent from the dispersion system. One way to remove solvent is to burn out organic materials during the sintering process. Organic removal must proceed with care. Sometimes care must be taken as the nanoparticles would lose their desired properties or activities during sintering. Nanoscale manufacturing may also involve traditional ceramic processing such as sintering and casting. Adair has extensive industrial collaboration through the NSF Particulate Materials Center at the Pennsylvania State University. Edirisinghe agreed that nano-sized powder processing is very difficult. These processes should be well-developed before nanomanufacturing so that we can understand nanomaterials processing better. He said that as evidenced at the symposium, nanotechnology research into the processing and forming of nanopowders is very limited worldwide.

TOP TEN TOPICS FOR ACADEMIC CERAMIC NANOMATERIALS RESEARCH

With so many scientists and researchers nowadays working on ceramic nanomaterials research, it is not easy to pick up top ten topics in this area. The six panelists had different views of the top ten topics in ceramic nanomaterials research. Adair picked dispersion as the top priority for his research. He emphasized the difficulties of handling and dispersing large volumes of nanomaterials as agglomerated powders. "Dispersion, dispersion, dispersion,..., and dispersion, this is my top ten topics of ceramic nanomaterials research," Adair repeated "dispersion" for ten times to underline his point of view. His targeted size of nanomaterials is below 50 nm. He believes that dispersion of nanomaterials is so important that agglomerated nanoparticle systems will lose the novel properties of nanoparticles, in some cases, without any practical applications. Edirisinghe explained that for structural materials applications, it needs a large quantity of nanomaterials. For functional and biological applications, both quantity and quality are very important. Furuya questioned the definition of nanomaterials. He mentioned that some people took materials about 1000 nm as nanomaterials. What is "nano" is very important. Each material has different aspects depending on their size. Size does matter for nanomaterials.

What is the definition of "nanotechnology"? While many definitions for nanotechnology exist, the National Nanotechnology Initiative (NNI) calls it "nanotechnology" only if it involves all of the following (see http://www.nsf.gov): (1) research and technology development at the atomic, molecular or macromolecular levels, in the length scale of approximately 1 - 100 nanometer range; (2) creating and using structures, devices and systems that have novel properties and functions because of their small and/or intermediate size; and (3) ability to control or manipulate on the atomic scale. Rogers picked up sol-gel hybrid organic/inorganic nanomaterials, biomaterials, low temperature growth and processing, colloidal nanoparticle systems, whiskers, nanowires, nanotubes, and nano-fabrication in the top-ten list. Wang listed shape control, ferrite nanoparticles, anistropoical nanostructures for magnetic application and data storage, semiconductor nanocrystals (doped and undoped), one-dimensional complex nanomaterials, PZT and barium titanate, functionalization of nanowires and nanobelts, biosensors, integrating nanostructures with MEMS and micron devices, and self-assembly and control growth. Webster picked up synthesis and applications of nanostructures, biomaterials, instruments for nanomaterials characterization, inorganic/organic interface, safety and health issue of nanomaterials, biological applications, devices for early detection, and nanotechnology for safety application. He emphasized that safety and health issue of nanomaterials is very critical for the future of nanotechnology. Overall, the top ten topics for academia nanomaterials research, spanning from synthesis, processing, and applications of nanomaterials, are: biomaterials and biological applications of nanomaterials, dispersion and processing of nanomaterials, nanowires/nanotubes/nanobelts/whiskers, safety and health issue of nanomaterials, nanomaterials synthesis, piezo- and ferro- nanomaterials, biosensors and catalysis, integrating nanostructures with MEMS and micron devices, nanofabrication, and instruments for nanomaterials characterization.

BIOLOGICAL INTERACTIONS OF NANOMATERIALS

As already discussed earlier, one of the top ten topics is the biological application of nanomaterials. "What is the biological interactions of nanomaterials?" asked a member of the audiences. This is also a common question from scientists and researchers who are in the field of biological applications of nanomaterials, who have interests in this field, or who have concerns about the applications of nanomaterials to human beings. Webster replied that there are some research activities going on in this field to study the biological interactions of nanomaterials. *In-vitro* studies are starting now. It will take a long time before we have any definite answers, however, the initial results are very promising as increased bone, vascular tissue, cartilage, and bladder tissue regeneration have been demonstrated on nanophase materials. European countries appear to be more focused on the issue of nanophase material safety issues than we are in US. We need much more emphasis on the health and safety aspects of nanomaterials. Webster said that information concerning the health consequences of manufacturing nanomaterials. In addition, if nanophase materials are used in the body through drug delivery or as implants, information on how the body and human cells deal with nanoparticules is badly needed. Webster warned that we should not jump into *in-vivo* testing before solving the problems of nanomaterials agglomeration and dispersion. Agglomeration of nanoparticles results in micron-sized particles, which may not penetrate cells. It takes many years of investigation and analysis to design proper nanomaterial systems for biological applications and to study how the body responds. On the other hand, Webster added, there is already a vaccine using nano-sized aluminum oxide, which was approved by FDA. It seems that aluminum oxide nanoparticles has no detrimental effect on human beings in this vaccine application.

THE FUTURE OF NANOTECHNOLOGY

While there is a "research heat wave" in the field of nanotechnology in recent years, many people are wondering about the future of nanotechnology. Will nanotechnology last long enough so that the society will be beneficial from fruitful researches in the current and next centuries? Will nanotechnology research diminish after the current "research heat wave"? What is the future of nanotechnology? In US alone, the government funding for nanotechnology research will gain a dramatic increase in the next few years. Large government funds are also allocated for nanotechnology in other countries such as Japan, China, and European countries. The number of publications and patents related to nanotechnology increases exponentially worldwide. Media coverage of nanomaterials has appeared more and more. On the other hand, warnings and concerns of negative health and environmental effects of nanomaterials have emerged in media, research publications, and scientific conferences. Adair pointed out that we have to make sure that in academia we do not oversell nanotechnology. We have seen overselling nanotechnology by media, giving the public an impression that nanotechnology can do everything. On the other hand, researchers and scientists cannot overstate their research and include everything as nanotechnology. This is related to the definition of nanotechnology as noted before. Many researchers claimed breakthroughs of nanotechnology even though the size of materials is in micrometer and sub-micrometer range. Nanotechnology deals with novel materials less than 100 nm with size induced novel functions. Wang agreed that the media and the public are the ones that have oversold nanotechnology.

Edirisinghe pointed out that the future of nanotechnology should be much better than any other technologies since it covers a broad remit. Rogers added that nanotechnology is much broader and will not go on the road of photonics and superconductors. Unlike other single disciplinary technologies, nanotechnology is multidisciplinary from the start. It involves multidisciplinary collaborations from physics, chemistry, mathematics, materials science, computer science, mechanical engineering, biology, medicine, and others. Because of its multidisciplinary nature, the future of nanotechnology, not relying on the boom and wane of a single technology, is much brighter. Furuya tried to distinguish microtechnology and nanotechnology. He stressed that nanotechnology is entirely at the atomic or molecular level, giving an opportunity of building nanomaterials atom by atom. Microtechnology should not be included as a part of nanotechnology. High-resolution transmission electron microscopy is a very powerful tool for observing nanocrystal formation and atom-to-atom building.

EXAMPLES OF NANOMATERIALS BEING PRODUCED

"What are the top five to ten oxides of interest as nanomaterials and what are the dimensions of their desired nanostructures?" was the last question from the audience. The panelists gave a list of nanomaterials and their corresponding desired sizes related their own researches and interests. Edirisinghe focused 20 nm silica and nano-hydroxyapatite. Webster mentioned TiO_2 and Al_2O_3 nanoparticles, and about 40 ~ 60 nm size of hydropyapatite. Adair is studying $BaTiO_3$ with a desired size of about 4 nm, ZrO_2 of about 8 nm, SiO_2 of about 10 to 20 nm, α-Al_2O_3 of about 40 nm, metallic nanomaterials of about 10 to 20 nm, and their aggregates. Furuya added nanobelts, nanowires, and boron nitride. Wang's research spans from ZnO, In_2O_3, Ga_2O_3, CdO, PbO_2 and SnO_2 nanobelts, nanowires, and whiskers, wide range of shapes. Typically, the as-synthesized oxide nanobelts are pure, structurally uniform, single crystalline and most of them free from defects and dislocations. They have a rectangular-like cross-section with typical widths of 30 to 300 nm, width-to-thickness ratios of 5 to10 and lengths of up to a few millimeters. The belt-like morphology appears to be a unique and common structural characteristic for the family of semiconducting oxides with cations of different valence states and materials of distinct crystallographic structures.

ACKNOWLEDGMENTS

The panel discussion "Nanotechnology: Past, Current, and Future" was a great success in terms of the panelists' expertise, knowledge, and high reputation in this field, the interesting topics during the panel discussion, and the audience's direct and active interactions with the panelists. Its format sets up a model for future discussions of emerging technologies with common interests. The symposium organizers thank the panelists for their participation, and gratefully acknowledge PPG Industries, Inc. for financial support of this symposium.

Author Index

Keyword Index